W9-BSC-905

MASONS
and
BUILDERS
LIBRARY

VOLUME I

by
LOUIS M. DEZETTEL

THEODORE AUDEL & CO.

a division of

HOWARD W. SAMS & CO., INC.

4300 West 62nd Street
Indianapolis, Indiana 46268

FIRST EDITION

FIFTH PRINTING—1977

Copyright © 1972 by Howard W. Sams & Co., Inc., Indianapolis, Indiana 46268. Printed in the United States of America.

All rights reserved. Reproduction or use, without express permission, of editorial or pictorial content, in any manner, is prohibited. No patent liability is assumed with respect to the use of the information contained herein.

International Standard Book Number: 0-672-23182-4

Library of Congress Catalog Card Number: 78-186134

Foreword

This volume on the fundamentals of concrete is a valuable reference book for those interested in any phase of concrete work. The information presented in this book shows you how to do your own repairs in concrete and tells you what you should know if you have a particular job to be done by a contractor.

Concrete is the most durable material available for construction. In addition to durability, concrete has great compressive strength. This permits it to carry its own weight, which is considerable, and allows unlimited types of structures to be built with concrete. Properly built concrete structures can, and have lasted for centuries.

Concrete is not a complicated science. The proportions of ingredients that go into the mixing of concrete are fairly exact. Anyone who will take the time to read and follow the information in this book can produce concrete construction of lasting quality, whether he is a professional contractor, a man who does occasional concrete work in connection with other construction work, or the handyman homeowner who likes to do his own work.

The author wishes to thank *Portland Cement Association* and its members who helped supply needed information in the preparation of this book.

Louis M. Dezettel

Contents

CHAPTER 1

Concrete

Concrete is one of the most economical and most important modern construction materials available. Because of its many advantages, it is commonly used in buildings, bridges, sewers, culverts, foundations, footings, piers, retaining walls, and pavements. Generally it is easily handled in its plastic state, and may be placed in forms and cast in almost any shape. Concrete has a tremendously long life and is extremely hard.

With all of the advantages of concrete it does have some limitations. Its tensile strength is not too great unless it is reinforced with steel bars or mesh. It is subject to some contraction and expansion under certain conditions of temperature and moisture. It has some slight permeability and will absorb water to a small degree.

Concrete is made up of a powdery substance called cement, water with which the cement chemically hydrates to form a hard rock-like mass, and stone aggregate used as a filler for economy. The term *portland cement* has become so much a common phrase that when we speak of cement or concrete we hardly give a thought to the meaning of the word "portland." It is not a trade name, nor a brand name, which is why it is not spelled with a capital P.

PORTLAND CEMENT

Portland cement was an English patent dating back to 1824, covering the manufacture of cement. Its color resembled a stone found on the Isle of Portland, so its inventor added the name to his cement. It has been known as portland cement from that time to the present.

Portland cement is a mixture of certain minerals which, when mixed with water, forms a gray-colored paste and cures into a very hard mass. The process of hardening is called *hydration* and is a chemical reaction between the water and the cement.

The principal ingredients of the cement mixture are obtained from limestone, rock, oyster and coquina shells, marl, shale, sand, iron ore, and clay. From these, lime, silica, and alumina, and iron components are obtained. They are finely pulverized and mixed together in certain proportions. The mixture is then fired in a kiln to about 2700°F and formed into clinkers. These clinkers, plus a small amount of gypsum (about 3%), are mixed together and ground up again. The gypsum controls the rate of hydration. This final ground-up fine powder is "portland cement."

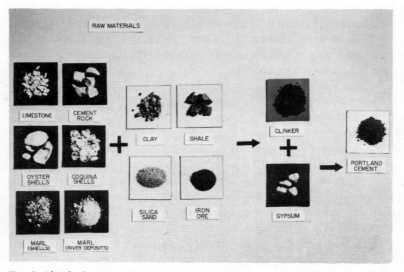

Fig. 1. The basic materials used to make portland cement which is the special ingredient of concrete.

The ingredients of portland cement are shown in Fig. 1. Figs. 2 through 5 are a flow chart showing the steps in its manufacture. Portland cement is produced to strict standards, which are set by

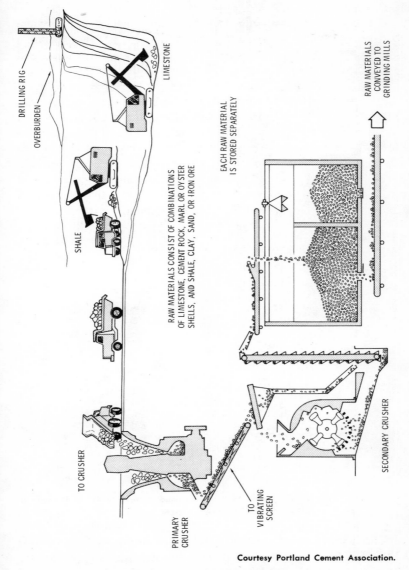

RAW MATERIALS CONVEYED TO GRINDING MILLS

DRILLING RIG

OVERBURDEN

LIMESTONE

SHALE

RAW MATERIALS CONSIST OF COMBINATIONS OF LIMESTONE, CEMENT ROCK, MARL OR OYSTER SHELLS, AND SHALE, CLAY, SAND, OR IRON ORE

EACH RAW MATERIAL IS STORED SEPARATELY

TO CRUSHER

PRIMARY CRUSHER

TO VIBRATING SCREEN

SECONDARY CRUSHER

Courtesy Portland Cement Association.

Fig. 2. Gathering raw materials for making cement.

11

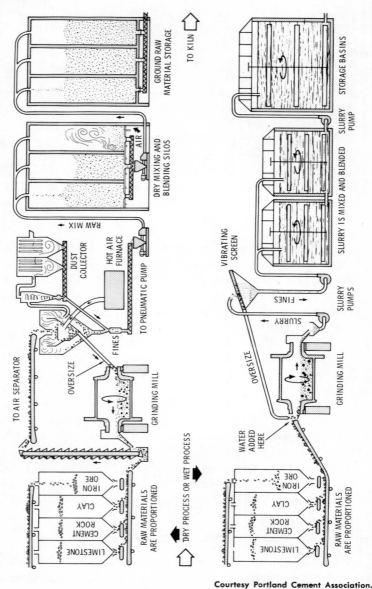

Courtesy Portland Cement Association.

Fig. 3. Wet or dry process of raw materials.

the *American Society for Testing and Materials* (ASTM). There are five basic types—each produced for specific purposes:

ASTM Type I
The general-purpose cement, used in all applications where special requirements are not specified. This is the cement supplied when a type number is not specified.

ASTM Type II
This type is used when precaution against moderate sulfate attack is needed. For example, in drainage structures where sulfate in ground water is present but not severe, Type II is recommended. Type II also generates less heat than Type 1 during hydration, which is helpful when pouring large masses of concrete.

ASTM Type III
This cement is also called "High-Early-Strength" cement. Concrete using this cement will cure earlier, and will have high strength in a week or less. It is recommended when forms must be removed promptly, structures must be put in service early, or where early winter curing is essential.

ASTM Type IV
A slow-curing cement where the heat generated by hydration must be kept to an absolute minimum. This is important in pouring large masses of concrete such as gravity dams.

ASTM Type V
For use where concrete is exposed to severe sulfate action. This may occur where concrete is in contact with soil or water with a high sulfate content. It is slow to gain strength compared with Type I or normal cement.

The *Canadian Standards Association* (CSA) has three standards — Normal, High-Early-Strength, and Sulfate Resisting. These match the ASTM Types I, III, and V.

AIR-ENTRAINING PORTLAND CEMENTS

By including small quantities of an air-entraining ingredient at the time of manufacture, concrete can be made to have millions of tiny air bubbles. This results in improved performance of concrete

13

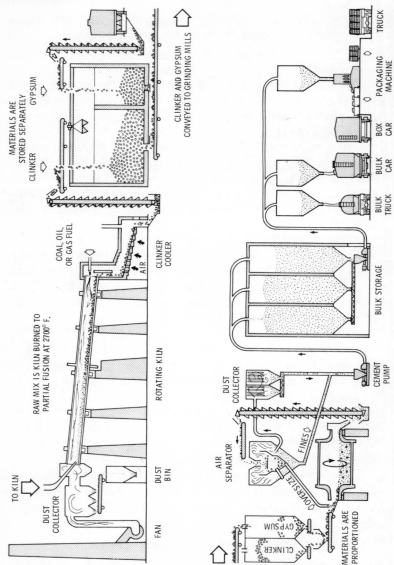

Courtesy Portland Cement Association.

Fig. 4. Processing raw material into cement clinkers.

Courtesy Portland Cement Association.

Fig. 5. Final stages in producing and shipping cement.

14

against the freeze-thaw action in northern climates, and longer life where chemicals are applied for snow and ice removal. Types I, II, and III cements are available with air-entraining ingredients, and are designated Types IA, IIA, and IIIA.

After three months of curing, the compressive strength of all five types is about equal. It varies considerably between types during the early days of curing. Table 1 compares the compressive strength of the five types as a percent of Type I or Normal.

Table 1. Approximate Relative Strength of Concrete as Affected by Type of Cement

Type of portland cement		Compressive strength, percent of strength of Type 1 or Normal portland cement concrete			
ASTM	CSA	1 day	7 days	28 days	3 mos.
I	Normal	100	100	100	100
II		75	85	90	100
III	High-Early-Strength	190	120	110	100
IV		55	55	75	100
V	Sulfate-Resistant	65	75	85	100

Courtesy Portland Cement Association.

SPECIAL CEMENTS

Variations to the above five types are available. The most useful are the following:

White Porcelain Cement

By using materials with a minimum of iron and manganese oxide a true portland cement is made that is pure white, compared to the gray appearance of regular portland cement. It is used purely for architectural purposes where white is important to the appearance. It is especially desirable when color is to be added, as in mortars and stuccos.

Portland Blast-Furnace Slag Cements

For general construction, where Type I cement is normally used, Type IS (or IS-A for air-entrained) may be substituted for economy. This cement includes the addition of selected blast-furnace slag in the grinding operation during manufacture.

Masonry Cement

Mortar for masonry must have good workability and plasticity and have a high water-retention ability. It is produced by adding entraining and other selected ingredients at the time of manufacture, meeting a fixed standard.

Portland-Pozzolan Cements

This is another type containing ingredients for economy purposes. In this, pozzolan (consisting of siliceous, or siliceous and aluminous material) is blended into the portland cement at the time of grinding. It is designated IP or IP-A. Properties are very much like Type I. This and blast-furnace cements are also used in mortars, as they have better nonstaining characteristics than straight Type I.

Other special cements are Oil-Well Cement, Waterproofed Portland Cement, and Plastic Cements. They have special additives to make them better adaptable to such special requirements.

AGGREGATE

To reduce cost, and because the overall results are better, natural rock pieces are added. The cement bonds or cements the rocks or stones together. This is concrete in its practical form.

The terms "pieces of rock" or "stones" are used here loosely. Uneven pieces or pieces all of one size would have many voids between them that would need to be filled with cement, and that too would be expensive. The answer is in the word *aggregate*. Aggregate means a quantity of assorted size stones, graduated from about 1½″ in diameter down to tiny pebble size. The maximum size depends on the quantity of concrete being poured and the thickness of the project. A ¾″ size is recommended as the maximum size for most home construction jobs. Aggregate purchased from a reliable supplier will have been well-graded to provide an assortment of sizes, the smaller stones filling the voids of the larger ones (Fig. 6).

Fig. 6. Well-graded stones with a minimum of dirt and other foreign matter.

Based purely on theory, it would be possible to consider an aggregate of perfectly round stones (marbles, for example) in which perfectly selected smaller marbles would fill the voids between the larger ones, and still smaller ones would fill the voids between the small marbles. If you carry the theory down fine enough, you could be left with no voids (and no room for cement), but this is only in theory.

By definition, an aggregate is the hard material, such as sand, gravel, stone, etc., that is mixed with cement and water to make concrete. Aggregates are of two kinds:

1. Fine.
2. Coarse.

A fine aggregate is one the particles of which will pass through a screen with ¼″ openings. For fine aggregate, the material commonly used is sand or fine quarry screenings of stone.

A coarse aggregate is one the particles of which will not go through a ¼″ screen and which are usually not larger than 1½″. Gravel or broken stone is the most common material for coarse aggregates. Sometimes pebbles are used.

17

CONCRETE

If only one grade (or size) of stone were used for a concrete mix, there would be many voids between the stones that would have to be filled with cement (actually the cement/water paste). This would be very uneconomical. Fig. 7 shows how two sizes of

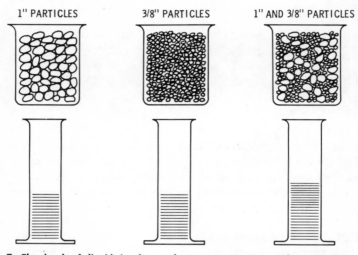

1" PARTICLES   3/8" PARTICLES   1" AND 3/8" PARTICLES

Fig. 7. The level of liquid in the graduates, representing voids, is constant if the particles are all the same size. When different sizes are combined, the void content decreases.

stones, separately, displace the same amount of water in a jar, and how the mixture of the same two sizes displaces a greater amount of water. The smaller size fills many of the voids between the larger size stones. Depending on how well graded the aggregate is, concrete may consist of from 60% to 80% aggregate.

Grading is determined by screen sieves of various mesh openings. For stones under ¼" (considered sand), the sieves are numbered from 4 to 100, based on the number of wires to the linear inch in the screen. Thus, a No. 4 sieve will pass stones ¼" in diameter or less, a No. 8, ⅛" or less, etc.

Grading screens for aggregate above the sand size are numbered in inches, from ¼" (equivalent to No. 4) to as much as 6". The usual standard sizes are, ¼", ⅜", ¾", 1", 1½", 3", and 6".

The larger the size of the maximum size aggregate in a graded load, the less cement and water is required. However, the cost of large sized aggregates can offset the added cost of cement and water, especially in sizes over 2″ or 2½″.

The recommended maximum size is affected by handling ability. As a rule of thumb the maximum size should not exceed:

1. One-fifth the dimension of nonreinforced members.
2. Three-fourths the space between reinforcing bars and surrounding forms.
3. One-third the depth of nonreinforced slabs on the ground.

Particle Shape

The shape of the aggregate particles has more effect on the mixing of fresh concrete than on the finished hardened concrete. There is no basic difference in strength due to the shape, but more cement and water material might be needed for crushed stone where the particles are elongated or slivery in shape. Odd-shaped particles should not exceed 15% of the total aggregate.

Skip-Graded Aggregate

When aggregate is dug from gravel banks and not screened and graded, it may contain an excess of fine aggregate or sand. This is called "skip-graded" aggregate. While it may be economical to buy, the economy is lost in the greater amount of cement required to fill the voids.

Sand

Sand is stone of various kinds, the same as for aggregates. However, those sizes ¼″ in size and smaller are classified as fine aggregate or sand. Sand is inexpensive and readily available.

Unlike the larger aggregate sizes, which may be purchased with well-spaced gradations, sand is normally not screened as well as larger aggregate. Thus, it will contain many sizes as it comes from its source or is ground from larger stones. Sand *mix* can be given a fineness figure, and this is known as the *fineness modulus*.

The *fineness modulus* (FM) figure is obtained by passing a sample load of sand through a number of sieves of increasing

19

fineness, and recording the percent of sand retained by each sieve size. The percentage figures are added, and then divided by 100. The final figure is the *fineness modulus*. The higher the figure, the coarser the sand. Table 2 shows an example of a test made on 1000 lbs. of sand. The FM figure of 2.80 shows that this sand is a good average sand for use in concrete.

Table 2. Fineness Modulus

| SCREEN SIZE | WEIGHT RETAINED | | CUMULATIVE % RETAINED |
	INDIVIDUAL	CUMULATIVE	
NO. 4	40	40	4.0
NO. 8	130	170	17.0
NO. 16	130	300	30.0
NO. 30	250	550	55.0
NO. 50	270	820	82.0
NO. 100	100	920	92.0
PAN	82	....	....
TOTAL WEIGHT	1000	....	280.0

$$\text{FINENESS MODULUS (FM)} = \frac{280}{100} = 2.80$$

NOTE: THE FINENESS MODULUS VALUES ARE
INTERPRETED AS FOLLOWS:

FINE SAND	2.20 — 2.60
MEDIUM SAND	2.60 — 2.90
COURSE SAND	2.90 — 3.20

The most desirable fineness of sand (fine aggregate) depends on the concrete desired for the particular job. Finer values are used in leaner mixtures, or when small-sized aggregates are used. In rich mixtures, coarser gradings are preferred for economy.

Lower FM values (coarser sand) may be used for easier placing of concrete, and where mechanical finishing is used. For hand finishing, a better looking finish is obtained when the sand is finer. The one-screen method may also be used for determining the sand application to a particular job. Most sands meet average use in

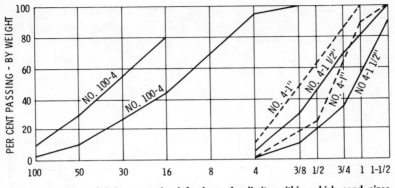

Fig. 8. The two solid lines on the left show the limits within which sand sizes should fall between the No. 4 and No. 100 sieve. The lines on the right are for coarse aggregates.

concrete if from 10% to 50% of the sand will pass the No. 50 screen. The 10% figure will produce a richer mix, and is for mechanical finishing. Sand passing from 15% to 30% through the No. 50 sieve, and even about 3% through the No. 100 sieve, is more ideal for hand finishing. The chart in Fig. 8 shows the acceptable limits of sand aggregates, and also for 1″ and 1½″ size coarse aggregate.

CLEANNESS

Any appreciable amount of silt, loam, clay, and other ingredients, affects the strength of the final concrete. The presence of these foreign ingredients reduces the ability of the cement/water paste to adhere to the aggregate particles and make a good bond. Suppliers will usually ship properly washed material. As a further precaution, it is very easy to turn a hose on a pile of coarse aggregate and wash it fairly clean (Fig. 9). Sand can be tested for the presence of excessive organic material or fine silt.

TESTING SAND

Up to 3% of the sand may pass the No. 100 sieve, but any that can pass the No. 200 sieve forms a silt or clay that affects

Fig. 9. Any size stone under ¼″ in diameter is called sand. Several sizes, and a mixture of all of them, are shown here. It is not important that the sand be so well graded—just so it is clean.

the quality of the concrete. Fine silt or clay coats the aggregate and prevents a good bond by the cement. A simple test for this is to fill a quart jar with dry sand to a depth of about 2″. Add water to about the ¾ mark. Shake well, then let stand for about an hour. A silt layer will settle at the top of the sand. If this layer is more than ⅛″ thick, the sand has too much silt.

Excessive organic material will weaken concrete to a small degree. Some organic materials (sugar, for example) will cause an increase in weakness as time passes. A test with caustic soda is used to examine sand for excessive organic matter. To test the sand, purchase 1 oz. of sodium hydroxide and a 12-oz. bottle from your local druggist. Dissolve the sodium hydroxide in 1 quart of distilled water. Put sand into the 12-oz. bottle to the 4½ oz. mark. Add the dissolved sodium hydroxide (now a 3% solution of caustic soda) to the 7 oz. mark. Cover with a rubber stopper, shake well, and let stand for 24 hours. Inspect the liquid for color. If it is clear, there is no organic matter in the sand. If it is a straw color, there is some organic material but not enough to affect the quality of the concrete. If the liquid is a dark color, there is too much organic material and the sand should be thoroughly washed or not used. Be careful in handling the sodium hydroxide. It can burn the skin, and is injurious to clothing. Wear rubber gloves while handling.

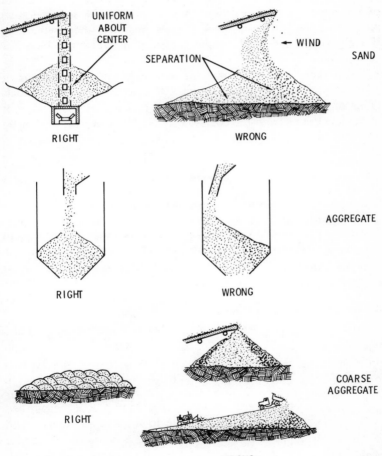

Fig. 10. Right way and wrong way of handling and storing sand and aggregate.

STORING

Certain precautions should be observed in storing aggregates and sand to assure consistent gradation of the aggregate and sand sizes. Storing in a cone-shaped pile should be avoided as there is a tendency for the material to slide down the sides and some segregation results. Ideally, the material should be stored in bins. The bins should have an opening at the bottom for the removal of the material. On large jobs, where yard piling is imperative, purchase the aggregate in separate sizes, and store them separately. When concrete is needed, proper grades of aggregate can be mixed at any time.

When sand is delivered wet, as is usually the case, there is no problem with separation due to wind. If delivered dry, avoid stock-piling on a windy day. Otherwise, there is tendency for separation of the fine grains from the coarse. Fig. 10 shows correct and incorrect methods of storing.

ADMIXTURES

It is seldom necessary to make additions to the ingredients of well proportioned concrete (see next chapter). If certain requirements must be met, it is best to select the cement type that meets those requirements in the first place, or alter the mixture to suit your particular needs. There are times, however, when admixtures are necessary.

Air-Entraining Admixtures

All cement types are available with air-entraining additivies, and may be ordered by adding the suffix A to the type of cement. If additional air-entraining is needed, or air-entraining is to be added to standard cement on hand, air entraining admixtures are available. It is added to the concrete at the time of mixing. Air-entraining adds to the ability of concrete to stand up under conditions of freezing and thawing, and reduces surface scaling as a result of salt or calcium chloride applied for snow and ice removal. It also improves the plasticity of the concrete for easier placement and finishing.

24

Water-Reducing Admixtures

Concrete strength may be increased if the water content can be reduced, and a given slump (described in the next chapter) is maintained. Water-reducing admixtures may retard the setting time and affect the timing schedule in handling large amounts of concrete at one time. Water-reducing admixtures with air-entraining material is also available.

Retarding Admixtures

When pouring large batches of concrete in summer, when temperatures are high, hydration takes place more rapidly. Under certain conditions this may affect the placement and handling. Initial hardening may be reduced by cooling the mixing water and/or the aggregate, or by the use of a retarding admixture. However, there is some loss of early strength which is generally up to 3 days.

Accelerating Admixtures

The use of Type III, or "High-Early-Strength," portland cement is the best way of getting accelerated setting time and early strength. Where it is necessary to secure quicker setting in selected batches of concrete, calcium chloride may be added as an accelerator.

Calcium chloride must be dissolved in the mixing water. The recommended amount is 2% by weight to the cement. If added to the dry ingredients, some of the calcium chloride may not dissolve, and the result may be "popouts" or dark spots in the concrete. If added in amounts greater than 2%, stiffening may come too early, and problems of placement may be faced.

A 2% solution of calcium chloride is too weak to affect reinforced concrete. However, it should not be used in prestressed concrete, where embedded aluminum is used, or if in contact with galvanized steel.

Calcium chloride should not be used as an antifreeze in winter. In the concentration mentioned, it would raise the freezing point only slightly. There are better methods for pouring concrete in cold or inclement weather. This is better described in Chapter 7.

Pozzolan Admixture

Pozzolan is a siliceous or siliceous and aluminous material, which, in finely divided form and in the presence of water, will chemically react with calcium hydroxide and will have cementitious properties. It eliminates or reduces expansion due to alkali-reactive aggregates where the use of such aggregates cannot be avoided. It is slow in hydrating and produces little heat.

Pozzolan is seldom used as a cement substitute, but some of its benefits may be realized by ordering cement types with the P suffix, or by adding pozzolan to the mix. Sometimes the slow release of heat due to hydration becomes a problem when pouring large masses of concrete, such as for dams. Lowering the temperature of the water and/or aggregate in the mix, using a cement type with the P suffix, and using a pozzolan admixture, are methods of reducing heat buildup. Sometimes all three methods are used. It must be remembered that the addition of pozzolan reduces the early-strength qualities of concrete.

Plasticizers

Under certain conditions workability is important. Mortar used for brick-laying, stucco, and concrete finishing by trowel call for improved workability. A common additive to mortar is the addition of a small amount of lime. It has good adhesive power, and it makes mortar more plastic than straight portland cement.

Other methods of improving the workability of concrete are the use of more cement in the mix, the use of finer sand, and adding air-entraining admixtures. Air-entraining acts like a lubricant and makes concrete easier to work.

WATER

Water, mixed with cement, results in hydration and the forming of a hard solid. This hard solid adheres to the aggregate material which makes up the bulk of concrete. The quality of the water is important to the concrete. In general, any water you can drink is fine for use in concrete. However, water with a high sulfate content, even though drinkable, should not be used. Sea water may also be used, although it reduces the compressive strength.

CHAPTER 2

Proportions and Mixing

Concrete is made of specified proportions of cement, water, sand, and aggregate. The proper proportion, or ratio of one ingredient to the other, depends on two things—the application for which the concrete is intended and the desired workability. There are several methods for determining the right proportions. The first method is to "go by the books," that is, by charts and tables established from experience. After reviewing the charts, a trial batch should be made up using one bag of cement, plus the charted amounts of the other ingredients. This method is important on large jobs where it should be known beforehand what the final results will be. The second method is the "slump test." This is a small sampling which should be made for each batch of sand and aggregate delivered because of the possibility of variations of their quality, especially of the sand. A third method which has some legitimacy is intuitive knowledge, based on years and years of experience.

It has been traditional to refer to a concrete mix in terms of the ratio of cement-to-sand-to-aggregate. For example, the ratio 1:2:4 means, 1 bag of cement (1 bag of cement has a volume of 1 cu. ft., and weighs 94 lbs.), 2 cu. ft. of sand, and 4 cu. ft. of aggregate. Unfortunately, such a ratio does not take into account the most important ingredient—water. It is important, therefore, to explore the cement-to-water ratio first.

PROPORTIONS AND MIXING

WATER-TO-CEMENT RATIOS

The ratio of water to cement should be *constant,* regardless of the amount of sand and aggregate used. But it becomes a *constant* only after establishing the ratio based on the consideration of a few factors. The factors are the desired strength of the concrete and the wetness of the sand.

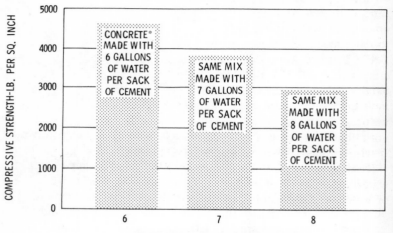

WATER-GALLONS PER SACK OF CEMENT

* TYPE 1 PORTLAND CEMENT, CONSTANT PROPORTIONS OF CEMENT, SAND AND COARSE AGGREGATE. MOIST CURED AT 70° F. FOR 28 DAYS.

Courtesy Portland Cement Association.

Fig. 1. The effect on final cure strength of concrete when the cement/water ratio is changed.

The less water used, the greater the strength of the cured concrete will be. The more water used, the easier it is to place (or work) the concrete. The choice of the amount of water should be based on the application. The effect of water on eventual concrete strength is clearly shown in the bar chart of Fig. 1. Note that when 6 gallons of water to a sack of cement are used, the 28-day compressive strength of the concrete will be 50% greater than if 8 gallons of water are used.

A more detailed chart is shown in Fig. 2. The shaded areas show the relative strengths of using water ratios of 4 gallons up to 8 gallons to a sack of cement, for various curing times from 1 day to 28 days. It is evident from these charts that a minimum of water results in a maximum of concrete strength, regardless of the curing time. Both standard (Type I) portland cement, and the "High-Early-Strength" (Type III) portland cement are shown.

Any reference to water amounts assumes *total water*. *It is based on the use of dry sand.* Generally, the supplier always wets the sand delivered to keep it from blowing away in the wind. The amount of water varies. However, a simple test can be applied to determine the wetness of the sand.

Fig. 3 show three degrees of wetness commonly used in concrete work. At (A) the sand is very wet. When you squeeze a handful of sand and open your hand, very wet sand will leave moisture on the palm of the hand. At (B) the sand has average wetness. Squeezing a handful of sand will leave a ball that does not fall apart when you open your hand. At (C) the sand is damp. It will fall apart when you open your hand.

Table 1 shows the effect of damp sand on the total amount of water needed. For example, when the formula calls for 6 gallons of water, and the sand is of average dampness (shown as wet in

Table 1. Water Usage

If mix calls for:	Use these amounts of mixing water, in gallons, when sand is:		
	Damp	Wet	Very wet
6 gal. per sack of cement	5½	5	4¼
7 gal. per sack of cement	6¼	5½	4¾

the chart), only 5 gallons of water are actually needed. The other gallon of water is in the sand.

Table 2 is more complete in that it includes the applications and suggests the water-cement ratio for those applications, as well as taking into account the wetness of the sand. Included are figures for making a trial batch from 1 sack of cement, and the amounts of ingredients needed to mix 1 cu. yd. of concrete.

Table 3 is a simplified chart showing the water to be used for average wetness sand, as well as the cement-sand-aggregate pro-

29

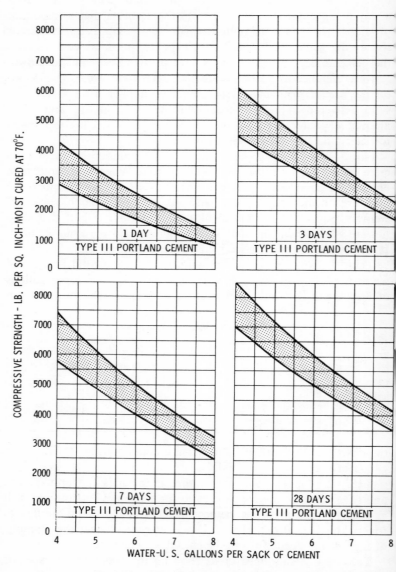

Fig. 2. The effect of cement/water ratios is shown in greater detail.

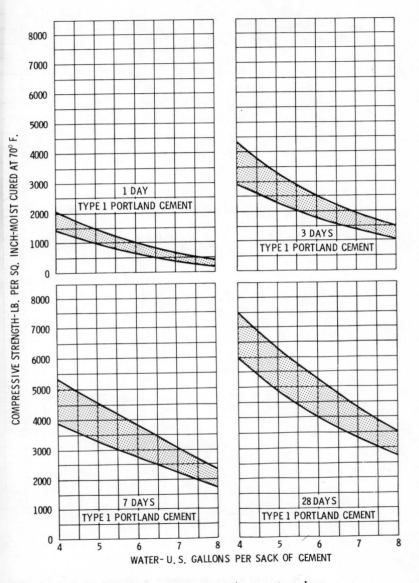

Compressive strength is almost in inverse proportion to water volume.

(A) Very wet sand leaves a large amount of moisture on the palm of your hand.

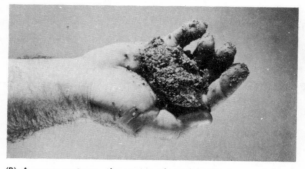

(B) Average wetness—the grains of sand hold up as a ball or large clump.

(C) Damp sand—it falls apart into small clumps.

Fig. 3. Method used for testing sand wetness.

Table 2. Recommended Proportions of Water to Cement

KINDS OF WORK	Add U. S. Gal. of Water to Each Sack Batch If Sand Is			Suggested Mixture for Trial Batch			Materials per cu. yd. of Concrete*		
	Very Wet	Wet	Damp	Cement Sacks	Fine Cu. Ft.	Coarse Cu. Ft.	Cement Sacks	Fine Cu. Ft.	Coarse Cu. Ft.
5-GALLON PASTE FOR CONCRETE SUBJECTED TO SEVERE WEAR, WEATHER OR WEAK ACID AND ALKALI SOLUTIONS									
Colored or plain topping for heavy wearing surfaces as in industrial plants and all other two-course work such as pavements, walks, tennis courts, residence floors, etc.	4¼	Average Sand 4½	4¾	1	1	1¾	10	10	17
						Maximum size Aggregate ⅜"			
One-course industrial, creamery and dairy plant floors and all other concrete in contact with weak acid or alkali solutions.	3¾	4	4½	1	1¾	2	8	14	16
						Maximum size Aggregate ¾"			
6-GALLON PASTE FOR CONCRETE TO BE WATERTIGHT OR SUBJECTED TO MODERATE WEAR AND WEATHER									
Watertight floors such as industrial plant, basement, dairy barn, etc. Watertight foundations. Concrete subjected to moderate wear or frost action such as driveways, walks, tennis courts, etc. All watertight concrete for swimming and wading pools, septic tanks, storage tanks, etc. All base course work such as floors, walks, etc. All reinforced concrete structural beams, columns, slabs, residence floors, etc.	4¼	Average Sand 5	5½	1	2¼	3	6¼	14	19
						Maximum size Aggregate 1½"			
7-GALLON PASTE FOR CONCRETE NOT SUBJECTED TO WEAR, WEATHER OR WATER									
Foundation walls, footings, mass concrete, etc., not subjected to weather, water pressure or other exposure.	4¾	Average Sand 5½	6¼	1	2¾	4	5	14	20
						Maximum size Aggregate 1½"			

portions recommended for flat slabs such as driveways and sidewalks. It is a good table for homeowners and small job contractors to use.

CEMENT-SAND-AGGREGATE PROPORTIONS

Having shown the importance of the water-cement ratio, we may return to the traditional ingredient proportion figures. As men-

Table 3. Recommended Proportions for Flat Concrete Work

	Portland cement (94-lb. sacks) (1 cu. ft.)*	Water for sand that is:				
		Very wet	Average wet	Damp	Sand	Aggregate
For 1 sack portland cement	1 sack	4¼ gals.	5 gals.	5½ gals.	2¼ cu. ft.	3 cu. ft.
For 1 cu. yd. concrete (27 cu. ft.)	6¼ sacks	25½ gals.	30 gals.	38 gals.	1400 lbs.	2100 lbs.

*One sack of cement makes about 4½ cu. ft. of concrete.

tioned before, a ratio of 1:2:4 means a mixture of concrete made up of:

1 part of cement.
2 parts of sand.
4 parts of stone.

Thus, in making up such a mixture, for every bag of cement, mix 2 cu. ft. of sand and 4 cu. ft. of stone or aggregate. In mixing concrete, satisfactory results are obtained by proper proportions.

The object of proportioning is to make a sufficiently dense concrete mixture. Sand, or gravel, or crushed stone alone, have between their particles empty spaces called "voids." To make dense concrete, the cement, sand, and stone must be proportioned so that the voids in the coarse aggregate are filled with the finer particles of sand and cement, and so that the voids in the sand are filled and bound together with the particles of cement. Table 4 shows this relation or shrinkage.

Table 4. Volume of Concrete Mixtures

Proportions of Mixture			Volume of concrete
Cement	Sand	Gravel or stone	
1 bag	1½ cu. ft.	3 cu. ft.	3½ cu. ft.
1 "	2 "	3 "	3⁹/₁₀ "
1 "	2 "	4 "	4½ "
1 "	2½ "	5 "	5⅖ "
1 "	3 "	5 "	5⅘ "

It is seen from Table 4 that the resulting volume of concrete is only a little more than the quantity of stone. For almost all work, twice as much gravel or stone must be used as sand. Concrete mixtures are classed according to the degree of richness due to the relative amount of cement used, as:

1. Rich.
2. Standard.
3. Medium.
4. Lean.

Rich Mixture—1 part cement; 2 parts sand; 3 parts coarse aggregate. Used for concrete roads and waterproof structures.

Standard Mixture—1 part cement; 2 parts sand; 4 parts coarse aggregate. Used for reinforced work floors, roofs, columns, arches, tanks, sewers, conduits, etc.

Medium Mixture—1 part cement; 2½ parts sand; 5 parts coarse aggregate. Used for foundations, walls, abutments, piers, etc.

Lean Mixture—1 part cement; 3 parts sand; 6 parts coarse aggregate. Used for all mass concrete work, large foundations, backing for stone masonry, etc.

How to Figure Quantities

The reason for shrinkage—that is, the ratio between the volume of concrete obtained and the volume of material used—should always be taken into account when calculating concrete. Thus, for example, a 1:2:4 mixture (in which 2 cu. yds. of sand and 4 cu. yds. of gravel are used) will not produce 6 cu. yds. of concrete because the sand will lodge in the spaces between the pebbles. If 6 cu. yds. of 1:2:4 concrete are wanted it will be necessary to use 2.64 cu. yds. of sand and 5.34 cu. yds. of gravel.

The quantities of cement, sand, and screened gravel required under average conditions for concrete of various proportions are shown in Table 5.

In estimating the amount of concrete required for a certain job, and the amount of materials required, either cu. ft. or cu. yd. may be used, although cu. yd. is a more convenient measurement particularly where large amounts of concrete are required. The following examples will serve to illustrate the method used in

35

Table 5. Approximate Quantities of Materials Required for Making One Cu. Yd. of Concrete

Proportions of the concrete or mortar			Quantities of materials		
Cement	Sand	Gravel or stone	Cement	Sand (damp and loose)	Gravel (loose)
			Sacks	Cubic yards	Cubic yard
1	1.5	....	15.5	.86	....
1	2.0	....	12.8	.95	....
1	2.5	....	11.0	1.02	....
1	3.0	....	9.6	1.07	....
1	1.5	3	7.6	.42	.85
1	2.0	2	8.2	.60	.60
1	2.0	3	7.0	.52	.78
1	2.0	4	6.0	.44	.89
1	2.5	3.5	5.9	.55	.77
1	2.5	4	5.6	.52	.83
1	2.5	5	5.0	.46	.92
1	3.0	5	4.6	.51	.85
1	3.0	6	4.2	.47	.94

(These quantities are approximate and may vary by 10% depending on the aggregate used.)

determining volume of concrete required for various types of work.

Example—A wall 9″ thick, 12 ft. high, and 30 ft. long has a door opening 3 ft. wide and 6 ft. high, also a footing 18″ wide and 9″ deep. The concrete is to be of the proportions 1:2:4. What quantities of material is required?

Solution—The volume of the footing is found by multiplying together the dimensions expressed in ft. Thus: $1\frac{1}{2}' \times \frac{3}{4}' \times 30'$ = 33.75 cu. ft. Similarly, the volume in the wall is $\frac{3}{4}' \times 12' \times 30'$, less the door opening $\frac{3}{4}' \times 3' \times 6'$ or $270' - 13.5' = 256.5'$ cu. ft. The total volume of the footing and wall is therefore $33.75 + 256.5$ or 290.95 cu. ft. Since there are 27 cu. ft. in 1 cu. yd. it follows that this sum should be divided by 27 to obtain the number of cu. yds. required. Therefore, $290.25/27 = 10.75$ cu. yds.

To find the necessary amount of cement, sand, and gravel, multiply the quantities for 1 cu. yd. as given in Table 5 (line 8) by 10¾, and it will be found that 64.5 sacks of cement, 4.73 cu. yds. of sand, and 9.57 cu. yds. of gravel is necessary to build the wall.

Example—It is required to build a two-course concrete sidewalk 27 ft. long and 4 ft. wide. The complete thickness of the walk is to be 6″, with a sub-base of 5″, and a wearing surface of 1″ thick. The sub-base is mixed in the proportions 1:3:5, whereas the wearing surface is mixed in the proportion 1:2. What quantities of material are required?

Solution—The volume of the sub-base is 27′ × 4′ × 5/12′ = 45 cu. ft. or 45/27 = 1⅔ cu. yds. The volume of the wearing surface is similarly 27′ × 4′ × 1/12′ = 9 cu. ft. or ⅓ cu. yd. Multiplying the quantities in line 12 of Table 5 by 1⅔ and those in line 2 by ⅓, it is found that the sub-base required 7.68 sacks of cement, 0.85 cu. yd. of sand, and 1.42 cu. yds. of gravel, while the wearing surface requires 4.27 sacks of cement, and 0.32 cu. yd. of sand.

Example—A circular tank 9 ft. inside diameter has walls 6″ thick, and is 4 ft. high above the floor. The floor is 6″ thick, and the concrete is to be 1:2:4. What quantities of material are required?

Solution—The calculation of a structure of this type can best be accomplished by considering the bottom and the cylindrical body separately. Thus, for the bottom the area is $5^2 \pi$ which multiplied by the height, gives a volume of $0.5 \times 5^2 \pi$ or 12.5π cu. ft.

The area of the cylindrical structure is $5^2 \pi - 4.5^2 \pi = 4.75 \pi$ sq. ft. The volume equals the area multiplied by the height which in this case is 3.5 ft. Therefore, the volume of the cylindrical structure is $3.5 \times 4.75 \times \pi$ cu. ft.

The total volume of the tank is therefore $12.5 \pi + 16.625 \pi = 29.125 \pi = 91.5$ cu. ft. This figure should be divided by 27 to obtain cu. yds. That is, $91.5/27 = 3.4$ cu. yds.

Multiplying the quantities in line 8 of Table 5 by 3.4, it will be found that the following material is required: 20.4 sacks of cement; 1.5 cu. yds. of sand; 3.1 cu. yds. of gravel.

The unit of measure for concrete is the cu. yd., which contains 27 cu. ft. To determine the amount of concrete needed, find the

Table 6. Ingredient
Proportions for Concrete

Maximum size of aggregate, inches	Water, gallon per sack of cement	Water, gallon per cu. yd. of concrete	Cement, sacks per cu. yd. of concrete	Fine aggregate —percent of total aggregate	Fine aggregate —lb. per sack of cement	Coarse aggregate—lb. per sack of cement	Fine aggregate —lb. per cu. yd. of concrete	Coarse aggregate —per cu. yd. of concrete	Yield, cu. ft. concrete per sack of cement
With Fine Sand—Fineness Modulus 2.20-2.60									
¾	5	38	7.6	43	170	230	1290	1750	3.56
1	5	37	7.4	38	160	255	1185	1890	3.65
1½	5	35	7.0	34	150	300	1050	2100	3.86
2	5	33	6.6	31	150	335	990	2210	4.09
¾	5½	38	6.9	44	195	250	1345	1725	3.91
1	5½	37	6.7	39	180	285	1205	1910	4.03
1½	5½	35	6.4	35	175	320	1120	2050	4.22
2	5½	33	6.0	32	175	370	1050	2220	4.50
¾	6	38	6.3	45	225	275	1420	1730	4.29
1	6	37	6.2	40	205	305	1270	1890	4.36
1½	6	35	5.8	36	200	355	1160	2060	4.66
2	6	33	5.5	33	200	400	1100	2200	4.91
¾	6½	38	5.9	46	245	288	1445	1700	4.58
1	6½	37	5.7	41	230	330	1310	1880	4.74
1½	6½	35	5.4	37	225	380	1215	2050	5.00
2	6½	33	5.1	34	225	430	1150	2195	5.30
¾	7	38	5.4	47	280	315	1510	1700	5.00
1	7	37	5.3	42	255	355	1350	1880	5.10
1½	7	35	5.0	38	250	410	1250	2050	5.40
2	7	33	4.7	35	250	465	1175	2185	5.75
¾	7½	38	5.1	48	300	330	1530	1680	5.30
1	7½	37	4.9	43	285	380	1400	1860	5.51
1½	7½	35	4.7	39	275	430	1290	2020	5.75
2	7½	33	4.4	36	275	495	1210	2180	6.14
¾	8	38	4.8	49	330	345	1585	1655	5.63
1	8	37	4.6	44	315	400	1450	1840	5.87
1½	8	35	4.4	40	305	455	1340	2000	6.14
2	8	33	4.1	37	310	525	1270	2150	6.59

Table 6. Ingredient
Proportions for Concrete (cont'd.)

Fine aggregate —percent of total aggregate	Fine aggregate —lb. per sack of cement	Coarse aggre-gate—lb. per sack of cement	Fine aggregate —lb. per cu yd. of concrete	Coarse aggregate —lb. per cu. yd. of concrete	Yield, cu. ft. concrete per sack of cement	Fine aggregate —percent of total aggregate	Fine aggregate —lb. per sack of cement	Coarse aggre-gate—lb. per sack of cement	Fine aggregate —lb per cu yd of concrete	Coarse aggregate —lb. per cu. yd. of concrete	Yield, cu. ft. concrete per sack of cement
With Medium Sand—		**Fineness Modulus 2.60-2.90**				**With Coarse Sand—**		**Fineness Modulus 2.90-3.20**			
45	180	220	1370	1670	3.56	47	185	210	1370	1595	3.56
40	165	250	1220	1850	3.65	42	175	240	1295	1775	3.65
36	160	290	1120	2030	3.86	38	170	280	1190	1960	3.86
33	160	325	1055	2140	4.09	35	170	315	1120	2080	4.09
46	205	240	1415	1655	3.91	48	215	230	1480	1585	3.91
41	190	275	1270	1840	4.03	43	200	265	1340	1775	4.03
37	185	315	1185	2015	4.22	39	195	305	1250	1950	4.22
34	185	360	1110	2160	4.50	36	195	350	1170	2100	4.50
47	235	265	1480	1670	4.29	49	245	255	1540	1610	4.29
42	215	295	1335	1830	4.36	44	225	285	1395	1770	4.36
38	210	345	1220	2000	4.66	40	225	335	1305	1945	4.66
35	210	390	1155	2145	4.91	37	220	380	1210	2090	4.91
48	255	280	1505	1650	4.58	50	265	265	1560	1560	4.58
43	240	320	1370	1825	4.74	45	250	310	1425	1765	4.74
39	235	370	1270	2000	5.00	41	250	355	1350	1920	5.00
36	235	415	1200	2120	5.30	38	250	405	1275	2065	5.30
49	290	305	1565	1650	5.00	51	300	290	1620	1565	5.00
44	270	340	1430	1800	5.10	46	280	330	1485	1750	5.10
40	265	395	1325	1975	5.40	42	270	385	1350	1925	5.40
37	265	450	1245	2120	5.75	39	280	435	1315	2045	5.75
50	315	315	1605	1605	5.30	52	330	300	1685	1530	5.30
45	300	365	1470	1790	5.51	47	310	355	1520	1740	5.51
41	290	415	1365	1950	5.75	43	305	400	1435	1880	5.75
38	290	480	1275	2110	6.14	40	305	465	1340	2045	6.14
51	345	330	1660	1585	5.63	53	360	315	1730	1510	5.63
46	330	385	1520	1770	5.87	48	345	370	1590	1700	5.87
42	320	440	1410	1935	6.14	44	335	425	1475	1870	6.14
39	325	510	1330	2090	6.59	41	340	490	1395	2010	6.59

volume in cu. ft. of the area to be concreted and divide this figure by 27. The following formula can be used to determine the amount of concrete needed for any square or rectangular area:

$$\frac{\text{Width (ft.)} \times \text{Length (ft.)} \times \text{Thickness (ft.)}^*}{27} = \text{cu. yds.}$$

For example, a 4″ thick floor for a 30 ft. × 90 ft. building would require:

$$\frac{30 \times 90 \times 0.33}{27} = 33.00 \text{ cu. yd. of concrete}$$

The amount of concrete determined by the above formula does not allow for waste or slight variations in concrete thickness. An additional 5% to 10% will be needed to cover waste and other unforeseen factors.

AGGREGATE SIZE

Ideally, aggregate should be well graded from ¼″ up. The maximum size depends on the size and shape of the concrete being poured, and the presence of steel reinforcements. The maximum size affects the handling ability. The maximum size should not exceed one-fifth the thickness of the concrete, nor three-fourths the space between steel reinforcements or to the edges of the forms. Ground-poured slabs should not have aggregate larger than one-third the thickness.

As a rule, the smaller the maximum size of aggregate, the greater the amount of water required. For improved strength, therefore, use the largest maximum sized aggregate that is economically possible, consistent with the recommendations mentioned above.

As a final chart for the selection of ingredients by the "book" method, consult Table 6. One column shows the amount of water and cement to use for different maximum size aggregate, as well as for differing water-to-cement ratios. Other columns show the lbs. of aggregate for sands of differing fineness.

---

*The thickness dimension must be changed to ft. or parts of a foot. The decimal part of a foot was used in the example. However, the fractional part may be used instead.

THE SLUMP TEST

A measure of the cohesiveness of the concrete mix is determined by the slump test. When large batches of concrete are being poured, the slump test should be made at intervals to assure some uniformity in the cohesiveness of the mix. The slump test is made by filling a form with concrete, removing the form, and measuring the amount of slump as the wet concrete settles down.

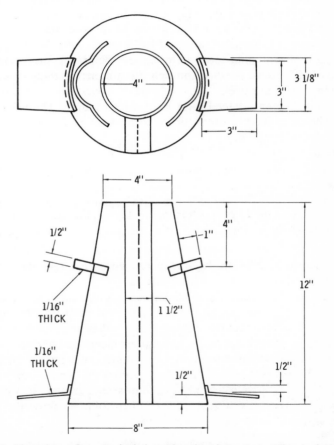

Fig. 4. Dimensions of a standard form for the slump test. Galvanized steel is generally used to make this form.

Fig. 4 shows the dimensions of the form. It is a tapered cylinder made of galvanized steel. Note the ears at the bottom and the handles near the top. A worker stands on the ears while the concrete is being rodded. The handles are for removing the form. In addition to the form, a tamping rod ⅝" in diameter and 24" long is required.

The form should be placed on a concrete floor or other nonwater absorbing slab. With a narrow scoop, fill the form with wet concrete in three layers. Each layer must be puddled with the tamping rod. Rod the first layer with 25 strokes of the rod. Scoop the second layer into the form and rod it 25 times, with the rod going through that layer and into the lower one. Fill the cone form and rod again 25 times, pushing through the intermediate layer. Strike off the top with a trowel.

Lift off the cone-shaped form right after rodding. The wet concrete should gradually slump down to a lower elevation. When it has stopped slumping, measure the distance between the top of the concrete and the top of the form. The amount of slump is the amount the concrete has fallen from the 12" height of the form, in inches.

A wet mix will slump more than a dry mix, and a dry mix may fall apart. A properly proportioned, workable mix will slump gradually, then hold its form. A 3" or 4" slump is considered normal.

Table 7. Allowable Slump for Various Concrete Applications

Types of construction	Slump, in inches	
	Maximum	Minimum
Reinforced foundation walls and footings	4	2
Unreinforced footings, caissons, and substructure walls	3	1
Reinforced slabs, beams, and walls	5	2
Building columns	5	3
Bridge decks	3	2
Pavements	2	1
Sidewalks, driveways, and slabs on ground	4	2
Heavy mass construction	2	1

Even a 6″ slump may be acceptable. The object of the test is not so much how much the slump is, but to be sure the slump is the same for all batches of concrete mixed to maintain uniformity of concrete for a job. Table 7 shows typical slump ranges and recommended limits in slump for various kinds of concrete construction.

TRIAL BATCH TEST

When pouring large amounts of concrete it is best to test its workability and ease of placement by first making a trial batch and inspecting it. Trial batches are generally based on the use of one sack of cement, plus the other ingredients. Use the "book" method first when mixing a trial batch. Tables 1, 2, and 3 show the recommended proportions to use for one sack of cement. It may be necessary to make several trial batches before a satisfactory mix is obtained.

Mix the trial batch in a rotating mixer for at least one minute. It is not necessary to mix for longer than three minutes. Remember, the water-to-cement ratio should remain constant. That ratio is a function of the application, is predetermined, and should not change. Measure the water accurately, using a 10-gallon bucket. Marks should be placed inside the bucket at each 2-gallon level.

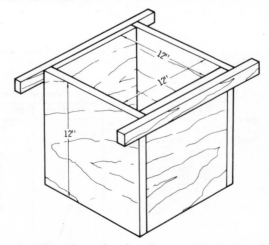

Fig. 5. Dimensions for a bottomless box for measuring one cu. ft. of aggregate.

(A) Too moist—add sand to this batch.

(B) Too dry—use less sand on the next batch.

(C) Troweled surface should be smooth and moist, but not wet.

Fig. 6. How to judge the proper stiffness of a concrete mix.

Use a china-marking or grease pencil. Changes should only be made in the amount of fine aggregate (sand) to coarse aggregate until the proper consistency is obtained.

Pour the mix from the mixer and inspect it for stiffness and workability. Make a slump test. If it is too wet, add sand and aggregate. If it is too stiff, reduce the amount of sand and aggregate. If it is too sandy or too stony, reduce the amount of sand or stone aggregate. It may take several trial batches to arrive at the right mix. Once determined, maintain the same proportions for the entire job, but do not make a slump test for each batch mixed.

For easy measurement of aggregates, make a bottomless wood box, similar to Fig. 5. Place the box on a flat slab and fill with sand or aggregate. Lift the box and the material will fall onto the slab. Each box load is 1 cu. ft.

For small jobs where one sack of cement is too much for a trial batch, smaller quantities may be used. Fig. 6 shows how to check a small, hand-mixed batch for stiffness. After mixing the proper proportions, smooth the top with a metal trowel. If the flat left by troweling has a layer of water that has oozed out of the batch, as shown in (A), add some sand and aggregate. If no moisture appears on the flat portion, as shown in (B), make a new batch with less sand. If the flat is smooth and moist, but not wet, as shown in (C), the batch is just right.

Table 6 should be convenient for industrial concrete contractors. It shows the ingredients suggested for various ratios of water-to-cement, and for various maximum sized aggregate content as well as for three grades of sand fineness.

HAND MIXING

Mixing 1 cu. yd. or more should always be done by machine. Hand mixing of large quantities is not economical. Hand mixing is done on a flat platform or concrete slab, or in a narrow mixing box (Fig. 7). Dry ingredients are first mixed by turning it three times, then water is added for another three-time turn.

If a tight floor is not available for mixing concrete, a watertight mixing platform should be made. This platform should be large enough for two men using shovels to work on it at one time. A

45

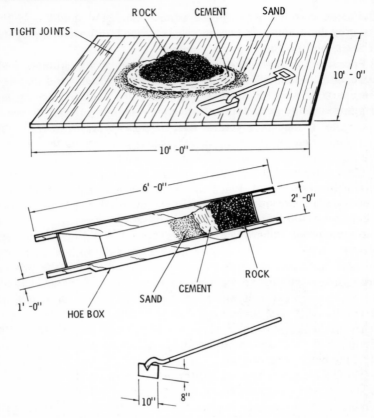

ROCK   CEMENT   SAND

TIGHT JOINTS

10' - 0"

10' -0"

6' -0"

2' -0"

ROCK

CEMENT

SAND

1' -0"

HOE BOX

10"   8"

Fig. 7. Simple equipment needed for hand-mixing small quantities.

good size is 10 ft. wide and 12 ft. long. It should be made pre-ferably of matched lumber so that the joints will be tight. Strips are nailed along the sides to prevent materials from being pushed off in mixing. The platform or mixing board should be well supported and set level prior to use.

The measured quantity of sand is spread out evenly on the plat-form and on this the required amount of cement is evenly distribu-ted. The cement and sand are turned with square-pointed shovels to produce a mass of uniform color, free from streaks of brown and gray. The required amount of coarse aggregate is then measured

46

and spread in a layer on top of the cement-sand mixture. Mixing is continued until the aggregate has been uniformly distributed throughout the mass.

A depression or hollow is then formed in the middle of the pile and the correct amount of water added while the materials are turned. This mixing is continued until the cement, sand, and aggregate have been thoroughly and uniformly combined.

POWER MIXING

For any amount of mixing using one sack of cement or more (which makes about 4½ cu. ft. of concrete), hand mixing is laborious and tiresome. With even this small amount, one is handling about 500 lbs. of material. It is prudent and more economical to use a power mixer.

Fig. 8. A small portable concrete mixer which uses an electric motor.

Power mixers are large drums with fins or blades inside. The drum rotates and the blades stir the contents. They are mounted to permit tilting for loading and unloading. Small drums are rotated by an electric or gasoline engine. Small mixers are usually on wheels so they may be hooked to the back of a truck for transporting to and from jobs (Fig. 8). Large mixers are truck mounted, and often operated from the truck engine.

When mixing on the job, all materials should be close at hand so loading can be done evenly and quickly. Sand and gravel should be in piles close to the mixing operation. Sacked cement should be near, but covered with a tarpaulin to keep it dry. Running water and a measuring pail should be nearby.

Place about 10% of the required water in the mixer first. Add dry ingredients of each type a little at a time, along with the additional measured water. The mixing drum should be rotating while loading. For a small mixer, the usual minimum mixing time is one minute when 1 cu. yd. of concrete material is loaded. Add 15 seconds of time for each ½ cu. yd. additional. Never load a mixer above its rated capacity.

READY-MIX

In cities and communities where the service is available, the best method of purchasing concrete ready to pour is by the *ready-mix* method (Fig. 9). Producers of this form of concrete may call it by other names, some of which are invented by the producer, but it is the method of delivery by truck with a revolving drum which mixes the concrete as it is being delivered. The *ready-mix* method is available to the homeowner in quantities of 1 yd. (the expression "1 yd." means 1 cu. yd.) or more, and to contractors in any quantity above that. Its popularity has reached a point where even the largest concrete jobs are now being handled by the *ready-mix* method, with trucks waiting in line to pour their load.

Ready-mix concrete may be ordered to your specifications. Suppliers will deliver a 5, 6, or 7-sack mix, which means 5, 6, or 7 sacks of cement to each yd. of concrete. Water quantity may also be specified. As explained previously, these ratios are dependent on the type of construction for which the concrete is intended.

Fig. 9. A large truck with ready-mix concrete.

The most important consideration in the use of *ready-mix* concrete is to be prepared to accept pouring of the concrete at the moment of delivery. This means all forms must be prepared in advance, and must be secure enough to take the pressure of the heavy concrete poured into them. Braces must be secure. The truck must have access to the forms for footings or ditches, or the area for slab pouring, or you must be prepared with wheelbarrows and help. Lay planks from the area where the truck will park, to the area to be poured, for easy wheeling.

Where concrete is to be poured onto the ground, as for a driveway, sidewalk, or patio, be sure the ground has been properly prepared in advance. Dig the ground down a few inches below the intended depth of the concrete, and pour stones or cinders to fill the difference. Water the ground thoroughly, and tamp it firm with a tamper. If you are doing a one-man job, it is well to have a friend or hired helper on hand to assist with the wheelbarrow and with leveling. Have a screed, or leveling board, on hand, ready to level out areas with the first pouring of the concrete. Pour concrete

49

into the farthest point first, and work toward the source, leveling as you go. Finishing can be done after the truck leaves.

If bolts are to be placed into the wet concrete, as for fastening to the 2" × 4" plate of a framed wall to go up later, have the bolts handy, and put them into position as soon as the concrete is poured, well before it begins to set. If a reinforcing steel net is to be included in a concrete slab, have it in place before the concrete is poured. The net can be lifted up to a midpoint in the concrete thickness immediately after pouring.

The driver of a ready-mix truck is usually on a schedule. Often he or his assistant will give you a hand in the pouring; but don't count on that: have help ready. If the truck is forced to stand by while you do some finishing to forms or ground, you will be charged for the extra time.

Ready-mix concrete is manufactured by three methods of mixing:

1. Central-mixed concrete is mixed completely in a stationary mixer and is delivered either in a truck agitator, a truck mixer operating at agitating speed, or a special nonagitating truck.
2. Shrink-mixed concrete is mixed partially in a stationary mixer and the mixing is completed in a truck mixer.
3. Truck-mixed concrete is mixed completely in a truck mixer.

Industry standards require that when a truck mixer is used for complete mixing, each batch of concrete must be mixed not less than 70 and no more than 100 revolutions of the drum or blades at the rate of rotation designated by the manufacturer as mixing speed. If the batch is at least ½ cu. yd. less than the maximum capacity or if the truck is used to complete partial mixing (shrink mixing), the minimum number of revolutions at mixing speed is 50. All revolutions after 100 should be done at a rate of rotation designated by the manufacturer as agitating speed. Agitating speed is usually about 2 to 6 rpm. Mixing speed is generally about 8 to 12 rpm, although some specifications permit a minimum of 4 rpm and a maximum peripheral velocity of the drum of 225 ft. per minute.

These standards also require that the concrete is to be delivered and discharged within 1½ hours, or before the drum has revolved 300 times after introduction of water to the cement and aggregates or the cement to the aggregates. Mixers and agitators should be operated within the limits of volume and speed of rotation designated by the equipment manufacturer.

REMIXING CONCRETE

Fresh concrete that is left standing tends to dry out and stiffen before the cement has hydrated to its intitial set. Such concrete may be used if, upon remixing, it becomes sufficiently plastic that it can be compacted in the forms. Under careful supervision, a small increment of water may be added to delayed batches providing the following conditions are met:

1. Maximum allowable water-cement ratio is not exceeded.
2. Maximum allowable slump is not exceeded.
3. Maximum allowable mixing and agitating time (or drum revolutions) are not exceeded.
4. Concrete is remixed for at least half the minimum required mixing time or number of revolutions.

Indiscriminate addition of water to make the concrete more fluid should not be allowed since this lowers the quality of the concrete. Remixed concrete may be expected to harden rapidly. Subsequently, concrete placed adjacent to or above remixed concrete may cause a cold joint.

Table 8 is a convenient guide for ordering ready-mixed concrete. The price will be affected by the amount of cement per yd., and the per-yd. price by the total amount of concrete ordered per load.

PREPARED MIXES

What we have said about mixing concrete is based on your purchase of the separate ingredients for projects requiring at least 6 cu. ft. of concrete (from one sack of portland cement). For small projects such as putting in a clothesline pole or for patching, you can buy a number of prepared dry mixes, already proportioned, and needing only the addition of water to the dry mix.

Table 8. Ordering Ready-Mix Concrete

KIND OF WORK	ORDER THE FOLLOWING**
Bond beams Chimney caps Lintels Reinforced concrete beams, girders, and other sections Reinforced concrete floors, roof slabs, and top courses Septic tanks	A mix containing at least 6½ sacks of portland cement per cu. yd. and a maximum of 6 gal. of water per sack of cement (using ¾" max. size aggregate).
Basement floors Curbs Driveways Entrance platforms and steps Garage floors Patio slabs Sidewalks Slabs on ground Stairs Swimming pools	A mix containing at least 6 sacks of portland cement per cu. yd. and a maximum of 6 gal. of water per sack of cement.
Footings Foundation walls Retaining walls	A mix containing at least 5 sacks of portland cement per cu. yd. and a maximum of 7 gal. of water per sack of cement.

**Order air-entrained concrete for all concrete exposed to freezing and thawing.

Sakrete is a brand name for various types of bag mixes. You can purchase an 80-lb. sack of sand mix for patching purposes; an 80 lb. sack of mortar mix for brick and block laying or tuckpointing; or a 90-lb. sack of concrete mix. They are also available in smaller bags. The proper ratios of all the dry ingredients are contained in each bag. You must still do a thorough job of blending the ingredients in each bag and adding water as instructed. For very small patch jobs, sand mixes are available from local hardware stores in packages as small as 5 lbs.

SPECIAL CEMENTS

There are two patching concretes of special formula that deserve mentioning, although there are no special instructions for the purchase of separate ingredients or in mixing. Latex concrete consists of a portland cement used with liquid latex instead of water. It is available in two forms—a can of cement powder which is accompanied by another can of the liquid latex; or as 5-, 10-, or 40-lb. sacks of dry latex (cement and dry latex) to which you only add water.

Epoxy concrete consists of an epoxy resin, a hardener, and sand. The resin and hardener are furnished separately; you mix them with the sand to make epoxy concrete. It is similar to the two-tube epoxy glues or cements you buy everywhere.

Both of these products cure to a very hard concrete and are ideal for small patch work because they adhere well to old concrete and may be feathered out to a layer as thin as 1/16". They are much too expensive for construction work.

A few additional tables are included at this point for convenience in estimating needs. Table 9 is useful for flat areas (slabs) and

Table 9. Ingredients Required for 100 Sq. Ft. of Concrete

Thick-ness of con-crete, in.	Amount of con-crete, cu. yd.	Proportions*								
		1:2:2¼ mix			1:2½:3½ mix			1:3:4 mix		
		Cement, sacks	Aggregate		Cement, sacks	Aggregate		Cement, sacks	Aggregate	
			Fine cu. ft.	Coarse, cu. ft.		Fine, cu. ft.	Coarse, cu. ft.		Fine, cu. ft.	Coarse, cu. ft.
3	0.92	7.1	14.3	16.1	5.5	13.8	19.3	4.6	13.8	18.4
4	1.24	9.6	19.2	21.7	7.4	18.6	26.0	6.2	18.6	24.8
5	1.56	12.1	24.2	27.3	9.4	23.4	32.8	7.8	23.4	31.2
6	1.85	14.3	28.7	32.4	11.1	27.8	38.9	9.3	27.8	37.0
8	2.46	19.1	38.1	43.0	14.8	36.9	51.7	12.3	36.9	49.3
10	3.08	23.9	47.7	53.9	18.5	46.2	64.7	15.4	46.2	61.6
12	3.70	28.7	57.3	64.7	22.2	55.5	77.7	18.5	55.5	74.0

*Quantities may vary 10% either way, depending on character of aggregate used. No allowance made in table for waste.

gives the amount of the various ingredients needed for various thicknesses of the slab. Table 10 is for determining needs for building concrete walls, either for form-poured concrete building walls or earth retaining walls. For the home mechanic who needs to mix small amounts for clothes pole foundations and similar projects, Table 11 will be convenient. It shows the solid ingredients needed for quantities of concrete from 100 cu. ft. down to 1 cu. ft.

Table 10. Materials for Concrete Walls

Wall 7 ft. high—Material needed for each 10 ft. length.

Thickness	1:2:4 Mixture			1:2½:5 Mixture			1:3:6 Mixture		
	Bags Cement	Cubic Feet Stone	Cubic Feet Sand	Bags Cement	Cubic Feet Sand	Cubic Feet Stone	Bags Cement	Cubic Feet Sand	Cubic Feet Stone
8 in.	10⅓	20⅗	41⅕	8⅕	21	42	7½	22⅖	44⅘
9 in.	11⅗	23⅘	46⅖	9⅖	23⅗	47⅖	8⅖	25⅕	50⅖
10 in.	12⅘	25⅗	51⅖	10½	26⅖	52½	9⅖	28	56
12 in.	15⅖	30⅘	61⅗	12⅗	31½	63	11⅕	33⅗	67⅕
18 in.	23	46⅕	92⅖	18⅘	47⅕	94⅖	16⅘	50⅖	100
Wall 8 ft. high—Material needed for each 10 ft. length.									
8 in.	11⅘	23½	47	9⅗	24	48	8½	25½	51⅕
9 in.	13⅕	26⅖	52⅘	10⅘	27	54	9⅗	28⅘	57⅗
10 in.	14⅗	29⅖	58½	12	30	60	10⅗	32	64
12 in.	17⅗	35⅕	70⅖	14⅖	36	72	12⅘	38⅖	76⅘
18 in.	26⅖	52⅘	106	21⅗	54	108	19⅕	57½	115
Wall 9 ft. high—Material needed for each 10 ft. length.									
8 in.	13⅕	26½	53	10⅘	27	54	9⅖	28⅘	57½
9 in.	14⅘	29⅗	59⅕	12⅕	30⅖	60⅘	10⅘	32⅖	64⅘
10 in.	16½	33	66	13½	33⅘	67½	12	36	72
12 in.	19⅘	39⅗	79⅕	16⅕	40½	81	14⅖	43⅕	86⅖
18 in.	29¾	59½	119	24⅖	60⅘	121	21⅗	64⅘	130

Material for each 10 ft. of length of Footings 1:3:6: Mixture

Size (height x width)	Cement Bags	Sand Cu. ft.	Stone Cu. ft.
6 in. x 12 in.	4/5	2-2/5	4-4/5
7 in. x 14 in.	1-1/8	3-3/8	6-3/4
8 in. x 16 in.	1-1/2	4-1/3	8-2/3
9 in. x 18 in.	1-4/5	5-1/3	10-3/5
10 in. x 20 in.	2-1/5	6-1/5	13
12 in. x 24 in.	3-1/6	9-1/3	18-3/5
15 in. x 30 in.	5	14-3/5	29-2/5

Table 11. Materials Required for Small Quantities of Concrete

Cubic feet of Concrete	1:1½:3 Mixture			1:2:3 Mixture			1:2:4 Mixture			1:2½:5 Mixture			1:3:6 Mixture		
	Bags Cement	Cu. Ft. Sand	Cu. Ft. Stone	Bags Cement	Cu. Ft. Sand	Cu. Ft. Stone	Bags Cement	Cu. Ft. Sand	Cu. Ft. Stone	Bags Cement	Cu. Ft. Sand	Cu. Ft. Stone	Bags Cement	Cu. Ft Sand	Cu. Ft. Stone
100	28	42	84	25⅘	51⅗	77⅖	22	44	88	18	45	90	16	48	96
90	25⅕	37⅘	75⅗	23⅕	46⅖	69⅗	19⅘	39⅗	79⅕	16⅕	40½	81	14⅖	43⅕	86⅖
80	22⅖	33⅗	67⅕	20⅔	41⅓	62	17⅗	37⅕	70⅖	14⅖	36	72	12⅘	38⅖	76⅘
70	19⅗	29⅖	58⅘	18	36	54	15⅖	30⅘	61⅗	12⅗	31½	63	11⅕	33⅗	67⅕
60	16⅘	25⅕	50⅖	15½	31	46½	13⅕	26⅖	52⅘	10⅘	27	54	9⅗	28⅘	57⅗
50	14	21	42	13	26	39	11	22	44	9	22½	45	8	24	48
40	11⅕	16⅘	33⅗	10⅓	20⅔	31	8⅘	17⅗	35⅕	7⅕	18	36	6⅖	19⅕	38⅖
30	8⅖	12⅗	25⅕	7¾	15½	23¼	6⅗	13⅕	26⅖	5⅖	13½	27	4⅘	14⅖	28⅘
20	5⅗	8⅖	16⅘	5⅕	10⅔	15⅗	4⅖	8⅘	17⅗	3⅗	9	18	3⅕	9⅗	19⅕
10	2⅘	4⅕	8⅖	2⅗	5⅕	7⅘	2⅕	4⅖	8⅘	1⅘	4½	9	1⅗	4⅘	9⅗
9	2½	3⅘	7⅗	2⅓	4⅔	7	2	4	8	1⅗	4	8	1½	4⅓	8⅔
8	2¼	3⅜	6¾	2	4⅕	6¼	1¾	3½	7	1⅖	3⅗	7⅕	1¼	3⅞	7¾
7	2	3	6	1⅘	3½	5⅖	1½	3	6	1¼	3⅛	6¼	1⅛	3⅜	6¾
6	1⅔	2½	5	1⅗	3⅕	4⅗	1⅓	2⅔	5⅓	1 1/10	2¾	5½	1	3	6
5	1⅖	2 1/10	4⅕	1⅓	2⅔	4	1 1/10	2⅕	4⅖	9/10	2¼	4½	⅘	2⅖	4⅘
4	1⅛	1¾	3⅜	1	2	3⅕	⅞	1¾	3½	7/10	1⅘	3⅗	⅝	2	4
3	⅘	1¼	2½	¾	1½	2⅓	⅔	1⅓	2⅔	½	1⅓	2⅔	½	1½	3
2	9/16	⅞	1 11/16	½	1	1½	7/16	⅞	1¾	⅓	9/10	1⅘	5/16	1	2
1	9/32	7/16	⅞	¼	½	¾	7/32	7/16	⅞	⅕	½	1	5/32	½	1

CHAPTER 3

Tools

The first tool needed is a method of measuring the right proportion of ingredients to be mixed for concrete. For small jobs the 1 cu. ft. bottomless box shown in Chapter 2 is convenient. If the inside is marked off in quarters, it may be used for measuring ingredients needed for as little as 1½ cu. ft. of concrete. Fig. 1 is the

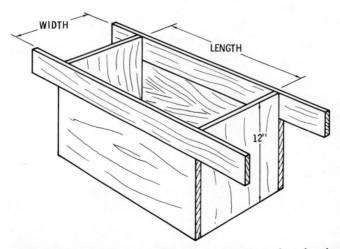

Fig. 1. Bottomless box and dimensions for various proportions of sand and coarse aggregate.

sketch of a bottomless box with a depth of 12″. Its length and width can be made to vary depending on the mix ratio. The table below the sketch shows the length and width dimensions for sand and stone quantities for various mix ratios. This is based on the use of one sack of cement, which is 1 cu. ft. Obviously, on jobs requiring many sacks of cement, the sand and stone (graded aggregate) proportions will be multiples of the quantities indicated from the bottomless box.

Experienced workers will soon learn the equivalent in shovels full, or wheelbarrows full, when feeding a power mixer. Inexperienced men must not be allowed to guess. They must measure the ingredients until they have gained the experience. Even with experience, a slump test should be made on each batch mixed. The most important ratio is the water/cement ratio, and requires that measured buckets of water be used. Do not run a hose into the mixer until it "seems" wet enough. Remember that excess water means weaker concrete.

Fig. 2. A gasoline-driven power mixer in use. Note the use of measured buckets of water.

CEMENT MIXERS

Mixers like that shown in Fig. 2 were described in the previous chapter. They vary in size and in the method of drive, from electric motors to gasoline engines.

Power mixers have their place for the homeowner and contractor where the timing will not match the delivery of *ready-mix* concrete, or in communities where *ready-mix* concrete service is not available. Wherever possible, the work should be scheduled to accept *ready-mix* trucks on the site, as this service is now the pre-

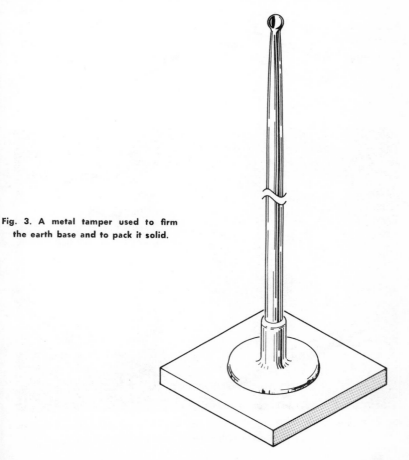

Fig. 3. A metal tamper used to firm the earth base and to pack it solid.

(A) General-purpose trowel.

(B) Pointing trowel.

(C) Wood float.

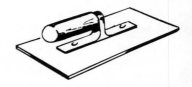

(D) Steel float.

(E) Edger.

(F) Strike board.

(G) Tamper.

(H) Broom.

Fig. 4. Sketches of basic tools used in concrete work.

Parts of Fig. 1 Chapter 3

Proportions of concrete	Box for sand	Box for gravel	Proportions of concrete	Box for sand	Box for gravel
	Ft.	Ft.		Ft.	Ft.
1:1:2.............	1 by 1	1 by 2	1:2½:5........	1¼ by 2	2 by 2½
1:1½:3............	1 by 1½	1 by 3	1:3:5............	1½ by 2	2 by 2½
1:2:4................	1 by 2	2 by 2	1:3:6............	1½ by 2	2 by 3

ferred method of purchasing and pouring concrete. The price of *ready-mix* concrete now competes with the purchase of the separate ingredients, and the need to purchase, store, and handle the ingredients is eliminated.

TAMPERS

Concrete for horizontal flat surfaces must be poured on a firm ground base. If it is not, cavities could form between the cured concrete and the ground beneath it, with the risk of the concrete breaking and dropping into the cavity. The ground is firmed by pounding it with a *tamper*. Fig. 3 shows a metal tamper. Illustration (G) in Fig. 4 is a tamper made from a piece of 4″ x 4″ lumber, with one end tapered for easy handling.

SCREEDS

Screed or *strike-off board,* the tool is the same by either name. Fig. 5 shows one in use. Illustration (F) of Fig. 4 is similar. A straightedge or strike-off rod is usually a straight piece of 2″ × 4″ or a 1″ x 4″ with a ½″ x 2″ shoe strip attached to the bottom. It can, however, be made from any straight piece of wood or metal that has sufficient rigidity. It is preferable to use a tool that has been specifically made as a straightedge instead of just a piece of lumber. The striking surface of a straightedge should always be straight and true. The straightedge should be longer than the widest distance between the screeds or edge forms. It is the first finishing tool used by the cement mason after the concrete is placed and is used to strike-off, or screed, the concrete surface to proper grade.

The strike-off board is used to scrape off the top of the concrete after first pouring it. It levels out the concrete as you draw it across, using the wood forms along the edge for a guide (Fig. 5). It usually takes two men to do the job right, one on each end of the board.

After the concrete has been struck off, hand tampers can be used to compact the concrete into a dense mass. They are used on flatwork construction with low-slump concrete. Such concrete is usually quite stiff and is often difficult to work.

Fig. 5. Screed, or strike-board, in use to level concrete slab.

One common form of hand tamper is better known in the trade as a "jitterbug." Its base is usually made of a metal grill 6½ " wide by 36" or 48" long (Fig. 6). Tampers are used to force the large particles of coarse aggregate slightly below the surface in order to enable the cement mason to put the desired finish on the concrete surface. This tool should be used only with concrete having a very low slump—about 1 "—and should be used to bring just enough mortar to the surface for proper finishing. The jitterbug or tamper should be used sparingly and in most cases is not recommended and not necessary, and the cement mason can proceed directly to *darbying,* the next step in quality finishing.

Fig. 6. A hand tamper, or jitterbug.

PUDDLING TOOL

The *puddling* tool, sometimes called a *spading* tool, is used to consolidate the concrete in poured forms. Fig. 7 shows the construction of one and how it is used. The puddling tool should be forced down along the inside of the forms and into the lower layer of poured concrete for several inches. Puddling should be continued until the coarse aggregate has disappeared into the concrete surface—its purpose is to provide even distribution of the coarse aggregate and mix the several layers of concrete poured into a form.

DARBY

A *darby* (Fig. 8) is a long, flat, rectangular piece of wood, aluminum, or magnesium from 30″ to 80″ long and from 3″ to 4″ wide with a handle on top. It is used to float the surface of the concrete

61

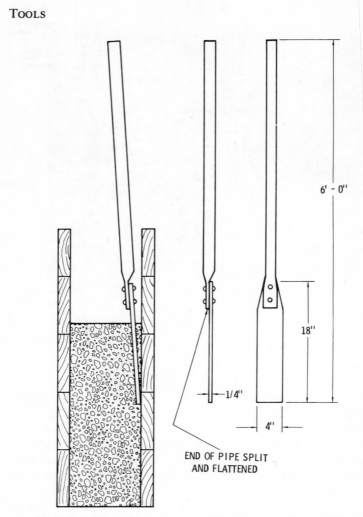

6' - 0"

18"

1/4"

4"

END OF PIPE SPLIT
AND FLATTENED

Fig. 7. A puddler tool is used to consolidate the concrete in poured forms.

slab immediately after it has been screeded to prepare it for the next step in finishing. This tool should be used to eliminate any high or low spots or ridges left by the straightedge. It should also sufficiently embed the coarse aggregate for subsequent floating and troweling.

Fig. 8. The darby is used to smooth the surface of concrete after screeding and before final finishing.

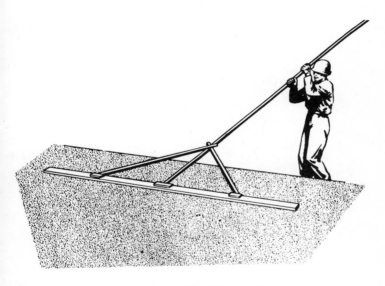

Fig. 9. Long-handled wood float.

FLOATS

After the surface moisture of the concrete slab has evaporated or been removed, and the concrete begins to take a set, *floats* are used to smooth the surface. Floats are made of wood or metal. A

63

wood float leaves a slightly textured surface and provides a little grip to the cured concrete for safer walking or driving. Metal floats result in a very smooth surface.

Fig. 9 shows a long-handled wood float used on large surfaces. The smoothing member is a piece of redwood or cypress, 4″ wide and as long as needed. Illustration (C) in Fig. 4 shows a small hand wood float, which permits the worker to get down on the concrete surface where he can better judge the effects of his work. The worker kneels on a large board on the concrete which, by the time it is ready for floating, is able to support his weight on the board. Illustration (D) in Fig. 4 shows a metal float. It is used in the same way as the wood float.

EDGERS AND GROOVERS

The *edger* is made of metal like the metal float, but it has a bent-over edge at right angles to the flat part (Fig. 10 and illustration

Fig. 10. An edger is used to round the edges of walks and other flat surfaces.

(E) of Fig. 4). It is used along the edge of a walk or slab to put a finished and smooth look to the concrete. It makes slightly rounded edges and a 4″ smooth border. A sharp edge could result in pieces of cured concrete breaking off later, which would have a poor appearance.

The *groover* is like the edger, but the vertical part is in the center (Fig. 11). When you pour a concrete walk or driveway, you must provide for the possibility of some settling of the base under

Fig. 11. The groover is used for cutting a crack line in large concrete slabs.

the slab, or some heaving due to freezing in the winter. The action would crack the concrete at irregular places, but when you score groove lines across the walk or driveway at regular intervals, the cracking will occur in the groove lines and will not show. The groover is used to put these indentations into the poured concrete about 3 to 5 hours after pouring, when the concrete has begun to cure and is hard enough for handling, yet soft enough for the groover to cut a line.

OTHER TOOLS

For surfaces that are even rougher than that made by a wood float, the hard-bristled broom (illustration (H) in Fig. 4) may be used after darbying. The small grooves left by the bristles make for good gripping, and are especially useful on driveways that are on an incline, such as a short approach to a garage.

Illustrations (A) and (B) of Fig. 4 show two trowels frequently used in concrete work as well as for bricklaying. Trowels are used in repair of concrete, for "buttering" pieces of concrete to cement, and for applying mortar to concrete blocks, where used.

POWER TROWELING

Contractors must consider the cost of labor in hand floating and troweling. To offset this, power troweling and floating is frequently used. Power trowels have three or four arms radiating from a

65

vertical shaft. Metal trowels are fastened to the arms, and are rotated in a circular motion.

After the concrete is stiff enough to support the weight of a man, the power float can be used to compact the concrete. The power trowel will give the surface a dense, smooth finish.

CURING MATERIALS

While there are many ways of preventing the rapid evaporation of water for proper curing, those requiring the purchase of special material will be mentioned here.

Waterproof Paper

Waterproof curing paper is an efficient means of curing horizontal surfaces and structural concrete of relatively simple shapes. One important advantage of this method is that periodic additions of water are not required. Curing paper assures suitable hydration of cement by preventing loss of moisture from the concrete. It should be applied as soon as the concrete has hardened sufficiently to prevent surface damage. The widest paper practical should be used. Edges of adjacent sheets should be over-lapped several inches and tightly sealed with sand, wood planks, pressure-sensitive tape, mastic, or glue.

Curing paper provides some protection to the concrete against damage from subsequent construction activity as well as protection from the direct sun. It should be light in color and nonstaining to the concrete. Paper with a white upper surface is preferable during hot weather.

Plastic Sheets

Certain plastic sheet materials are used to cure concrete. They are lightweight, effective moisture barriers and easily applied to simple as well as complex shapes.

In some cases, the use of thin plastic sheets for curing may discolor hardened concrete. This may be especially true when the concrete surface has been steel-troweled to a hard finish. When such discoloration is objectionable, some other curing method is advisable.

Curing Compounds

Liquid membrane-forming curing compounds retard or prevent evaporation of moisture from the concrete. They are suitable not only for curing fresh concrete, but also for further curing of concrete after removal of forms or after initial moist curing.

Curing compounds are of four general types—clear or translucent; white pigmented; light-gray pigmented; and black. Clear or translucent compounds may contain a fugitive dye which fades out soon after application. This helps assure complete coverage of the exposed concrete surface. During hot, sunny days, white-pigmented compounds are most effective since they reflect the sun's rays, thereby reducing the concrete temperature.

Curing compounds are applied by hand-operated or power-driven spray equipment immediately after the disappearance of the water sheen and the final finishing of the concrete.

CHAPTER 4

Concrete Slabs, Walks, and Driveways

Successful placing of concrete on flat ground requires considerable preparation in advance. This preparation includes:

1. Determine the slope of the land and correcting it if necessary.
2. Remove the earth to the depth required for the thickness of the concrete.
3. Firm the subgrade to prevent subsequent sinking.
4. Place edge forms or screeds.
5. Place joints to allow for expansion.
6. Correct pouring of the concrete.
7. Level the concrete preparatory to finishing.

USING A TRANSIT OR LEVEL

Large sites, such as home development areas, must be staked out and graded to provide level individual homesites for pouring horizontal concrete slabs. Before bulldozers can be ordered in, grade stakes must be placed. This can be done with a builder's level or transit.

Establishing Elevations With a Builder's Level

Set up the instrument where locations for which elevations are to be determined may be seen through the telescope. Level up the instrument and take a reading on the measuring rod by means of the horizontal cross hair in the telescope. The rod is then moved to the second point to be established. Then the rod is raised or lowered until the reading is the same as the original. The bottom of the rod is then at the same elevation as the original point.

Measuring Difference in Elevation

To obtain the difference in elevation between two points, such as (A) and (B) in Fig. 1, set up and level the instrument at an intermediate point (C). With the measuring rod on point (A) note the reading where the horizontal cross hair in the telescope crosses the graduation marks on the rod. Then with the rod held on point (B), sight on the rod and note where the horizontal cross hair cuts

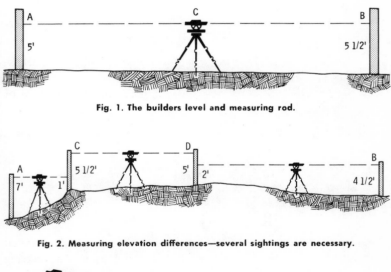

Fig. 1. The builders level and measuring rod.

Fig. 2. Measuring elevation differences—several sightings are necessary.

Fig. 3. Transit being used to establish points along a given line.

(A) Sighting transit on measuring rod.

(B) Sighting measuring rod for placing concrete forms.

(C) Marking proper height for concrete form.

Fig. 4. Steps in using a transit to

(D) Placing string line at proper height for form.

(E) Checking string line for proper subgrade depth.

(F) Placing concrete forms using string line as a guide.

set forms in a laterally straight line.

71

the graduations on the rod. The difference between the reading at point (A) (5 ft.) and the reading at point (B) (5½ ft.) is the difference in elevation between points (A) and (B). Thus point (B) is ½ ft. lower than point (A).

When, for any reason such as irregularity of the ground or a large difference in elevation, the two points whose difference in elevation is to be determined cannot be sighted from a single point, intermediate points must be used for setting up the instrument, as shown in Fig. 2.

Establishing Points on a Line With a Transit

Level the instrument and center it accurately over a point on the line by means of a plumb bob. Then sight the telescope on the most distant visible known point of that line. Lock the horizontal motion clamp screw to keep the telescope on line and place the vertical cross hair exactly on the distant point with the tangent screw. Then, by rotating the telescope in the vertical plane, the exact location of any number of stakes on that same line may be determined (Fig. 3).

PREPARING THE SITE

Before placing any concrete, the forms or screeds must be set to proper grade. The grade is sometimes determined by using a builder's level. The level is sighted on a measuring rod or rule that is set on an established grade (Fig. 4A) and a reading is taken. Then the rod or rule is moved to where the form or screed will be placed (Fig. 4B). With a stake in the ground at this point the rod or rule is raised or lowered until the desired grade is read by the level. A mark or nail is placed on the stake at the bottom of the rule (Fig. 4C). This operation is continued until enough stakes are set and marked with the proper grade and alignment.

A string line is then strung tightly from stake to stake at the marked position (Fig. 4D). This string line will be at the top of the form or screed. The distance from the string line to the sub-grade is checked to make sure there is enough depth to place the form or screed (Fig. 4E). If not, the subgrade is dug out until there is enough clearance. The forms or screeds are then set to proper line and grade by following the string line (Fig. 4F).

The forms or screeds are then well staked and braced. The stakes must be driven straight to ensure that the forms or screeds will be true and plumb. When the grade of a narrow walk is being established it is sometimes more convenient to set one edge form and then use a spirit level, as shown in Fig. 5, to set the other edge form.

Fig. 5. With one concrete form in place, the other side is constructed by using a spirit level.

All sod and vegetable matter must be removed from the construction site (Fig. 6) and any soft or mucky places must be dug out, filled with at least 2″ of granular material such as sand, gravel or crushed stone, and thoroughly tamped (Fig. 7). Exceptionally hard compact spots must be loosened and then tamped to provide the same uniform support for the slab as the remainder of the subgrade.

SAND BASE

If the soil is porous and has good drainage, concrete may be poured directly on the earth, if the earth is well tamped. If the

73

Fig. 6. Removing sod or any plant growth from construction site.

soil has a lot of clay and drainage is poor, it would be well to put down a thin layer of sand or gravel on which to pour the concrete (Fig. 8).

A short time before pouring, give the soil a light water sprinkling (Fig. 9). Avoid developing puddles; when the earth is clear of excess water you can begin pouring concrete. When additional fills are required under walks, driveways, or floors to bring them to the proper grade, the fills should also be of a granular material thoroughly compacted in a maximum of 4″ layers. It is best to extend the top of all fills at least 1 ft. beyond the edges of walks and drives, and to make the slope of the fill flat enough to prevent undercutting during rains.

Fig. 7. Tamping soil firmly.

The top 6″ of the subgrade should be sand, gravel, or crushed stone where subgrades are water soaked most of the time. These granular subgrades must be drained to prevent the collection of water. Well-compacted, well-drained subgrades do not require such special granular treatment.

EXPANSION JOINTS

Where a new concrete slab abuts an existing walk, driveway, building, curb, lighting standard, fireplug, or other rigid object, a premolded material, usually ½″ thick, should be placed at the joint. These joints are commonly called *expansion joints*. They are placed on all four sides of the square formed by the intersection of two walks. When the sidewalk fills the space between the curb and a building or wall, an expansion joint should be placed between the sidewalk and the curb and between the sidewalk and the build-

75

Fig. 8. Spreading a thin layer of sand or aggregate as a base for concrete.

Fig. 9. Sprinkling before pouring concrete will settle dust and compact soil.

ing or wall. Expansion joints are not required at regular intervals in the sidewalk.

PREPARING THE MIX

The concrete should contain only enough water to produce a concrete that has a relatively stiff consistency, works readily, and does not separate. Concrete should have a slump of about 3″ when tested with a standard slump cone. The adding of more mixing water to produce a higher slump than specified lessens the durability and reduces the strength of the concrete.

In northern climates where flat concrete surfaces are subjected to freezing and thawing, air-entrained concrete is necessary. Air-entrained concrete is made by using an air-entraining portland cement or by adding an air-entrained agent during mixing. Before the concrete is placed, the subgrade should be thoroughly dampened so that it is moist throughout, but without puddles of water.

Concrete should be placed between the forms or screeds as near to its final position as practicable. Precautions should be taken not to overwork the concrete while it is still plastic because an excess of water and fine material will be brought to the surface, which may lead to scaling or dusting later on. The concrete should be thoroughly spaded along the forms or screeds to eliminate voids or honeycombs at the edges.

FORMS

Edge forms may be of wood or metal. Contractors would do well to purchase metal forms designed for concrete work and which may be used again and again, indefinitely. Wood forms are usually 2″ × 4″, 2″ × 6″, or wider depending on the thickness of the concrete. Chapter 6 is devoted exclusively to forms for concrete.

Forms must be placed carefully, as the tops of the forms become screed guides for leveling the concrete. Distances apart must be measured accurately and a spirit level used to assure that they are horizontal. However, if the forms are used on an inclined driveway, they must follow the incline. Forms for curved walks or driveways may be made from ½″ redwood. When using redwood, soak it

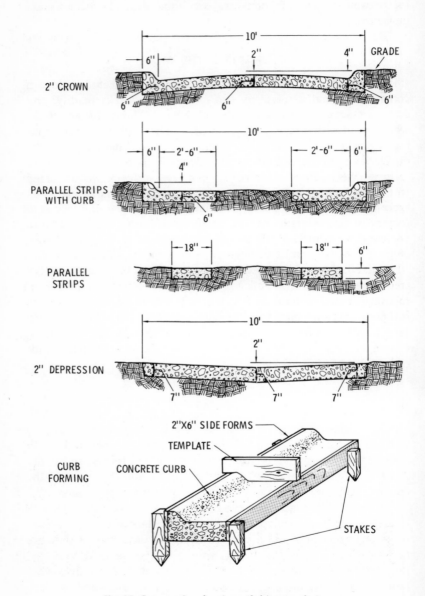

Fig. 10. Cross-section drawings of driveway designs.

with water for about 20 minutes, then it may be easily bent without splitting.

Stakes are placed at intervals along the outside of the forms and driven into the earth, then nailed to the forms to hold them securely in place. The tops of the stakes must be slightly below the edge of the forms so they will not interfere with the use of the strike-off board for screeding later.

DRIVEWAYS

Driveway construction is varied depending on the owner's preference. While most are flat surfaces with a width great enough for a single car, or for two cars such as approaches to a two-car garage, some are crowned, some depressed, and some consist of concrete runways laid parallel which allows for grass in between. A few driveways will be below grade level, with curbs at the edges. Fig. 10 shows cross sections of representative driveways other than the flat slab. The crowned and center-depressed driveway will require making a strike-off board with the proper shape. Fig. 11 shows the details of a driveway with a 1″ crown, and the strike-off board used for floating. The crown in the center of the driveway

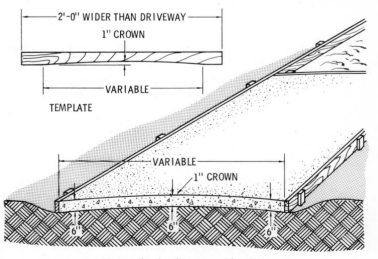

Fig. 11. Details of a driveway with a 1″ crown.

permits the run-off of water, especially important as ice or snow melts off the driveway.

Fig. 12 shows a cross section of a level driveway, when preferred, especially since a level slab is easier to screed and finish. Concrete highways are built perfectly level except on curves. It is important on high-speed roads that there be no pull to the side on the car as a result of a tilt in the road. This reduces driver fatigue on straight runs.

SIDEWALKS

Sidewalk construction is like that of driveways, except no curbs are used of the type shown and described above. Figs. 13 and 14 show the construction of sidewalks, which include expansion joints.

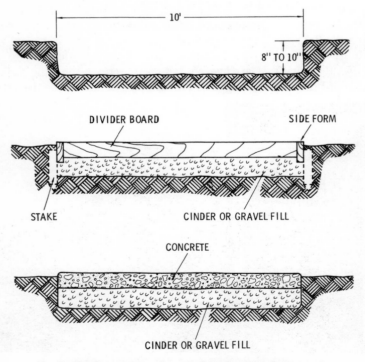

Fig. 12. Cross section of a typical single-car level driveway.

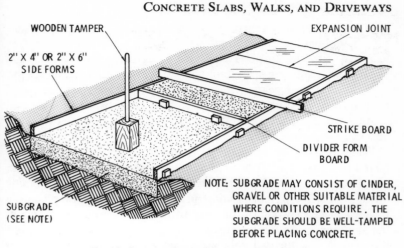

WOODEN TAMPER

EXPANSION JOINT

2" X 4" OR 2" X 6"
SIDE FORMS

STRIKE BOARD

DIVIDER FORM
BOARD

SUBGRADE
(SEE NOTE)

NOTE: SUBGRADE MAY CONSIST OF CINDER,
GRAVEL OR OTHER SUITABLE MATERIAL
WHERE CONDITIONS REQUIRE. THE
SUBGRADE SHOULD BE WELL-TAMPED
BEFORE PLACING CONCRETE.

Fig. 13. Steps in constructing a concrete sidewalk.

WHERE TO PUT JOINTS

Concrete expands and contracts slightly due to temperature differences. It may also shrink as it hardens. Joints should be put in the concrete to control expansion, contraction and shrinkage.

Fig. 14. This sidewalk width requires an expansion joint in the center.

It is desirable to prevent slabs-on-ground, either inside or outside, from bonding to the building walls. Thus, the slab will be free to move with the earth. To prevent bonding, a continuous rigid waterproof insulation strip, building paper, polyethylene, or similar material is placed next to the wall. These materials are also used next to other existing improvements, such as curbs, driveways, and feeding floors. The continuous rigid waterproof insulation strip acts as an isolation joint.

Wide areas, such as floor slabs and feeding floors, should be paved in 10 ft. to 15 ft. wide alternate strips. A *construction joint,* also known as a *key joint,* is placed longitudinally along each side of the first strips paved. A construction joint is made by placing a beveled piece of wood on the side forms. This creates a groove in the slab edges. As the intermediate strips are paved, concrete fills this groove, and the two slabs are keyed together. This type of joint keeps the slab surfaces even and transfers the load from one slab to the other when equipment is driven on the slabs.

Contraction joints, often called *dummy joints,* are cut across

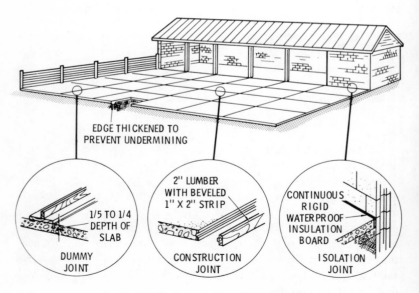

EDGE THICKENED TO
PREVENT UNDERMINING

1/5 TO 1/4
DEPTH OF
SLAB

DUMMY
JOINT

2" LUMBER
WITH BEVELED
1" X 2" STRIP

CONSTRUCTION
JOINT

CONTINUOUS
RIGID
WATERPROOF
INSULATION
BOARD

ISOLATION
JOINT

Fig. 15. Details of contraction and expansion joints.

each strip to control cracking. They do not extend completely through the slab, but are cut to a depth of one-fifth to one-fourth the thickness of the slab, thus, making the slab weaker at this point. If the concrete cracks due to shrinkage or thermal contraction, the crack usually occurs at this weakened section (Fig. 15).

Contraction joints should be cut soon after the concrete has been placed in order to work the larger pieces of coarse aggregate away from the joint. A simple way to cut this type of joint is to lay a board across the fresh concrete and cut the joint to the proper depth with a spade, axe, or similar tool. A groover is then used to finish the joint. Contraction joints are generally placed 10 ft. to 15 ft. apart on floor slabs, driveways, and feeding floors. They are placed 4 ft. to 5 ft. apart on sidewalks. All open edges should be finished with an edger to round off the edge of the concrete slab to prevent spalling.

Since the development of special blades that can saw concrete, the practice of sawing contraction joints is becoming common. The proper time to saw joints is generally 18 to 24 hours after the concrete has been placed. A sawed joint is clean and attractive and works very well when cut to the proper depth. On larger jobs, a special mobile concrete saw is used. On smaller jobs, such as sidewalks and driveways, a portable electric hand saw has been used with success. The operator must be careful to keep the saw blade straight so that it will not shatter.

POURING THE CONCRETE

Concrete should be poured within 45 minutes of the time it was mixed; otherwise, some curing begins to take place, and the concrete may become too thick to handle with ease. If you are pouring a large area, mix only as much as you can handle within the 45 minutes. In fact, the first batch should be a small one for trial purposes. After the first batch, you can determine whether the succeeding batches require more or less sand. *Remember,* do not vary the water for a thicker or thinner concrete—only the amount of sand. Once you have established the workability you like, stay with that formula.

83

Pour in the concrete up to the level of the form edges. Immediately after the first batch is poured into place, spade the concrete with an old garden rake or hoe to even it out approximately to the level of the form boards, and to make sure there are no air voids in the mix.

Fig. 16. After pouring, level concrete with a strike-off board.

Level off the concrete with the striker board mentioned in the previous chapter (Fig. 16). It frequently takes two men to do this —one at each end of the board. The object is to get a level and even top surface to the poured concrete using the board forms as the guide. Draw the striker board across the concrete while hugging the form edges; seesaw the board as you move it across. Now you can see why the stakes for the forms had to be below the edges of the form boards; they must not interfere with the movement of the strike board. The strike board takes off high spots and levels the concrete. If the board skips over pockets of low concrete, fill them in and go over it once more.

When the concrete is first poured and struck, there may be a sheen of water on the top, but not necessarily evenly across the top. You must wait until this sheen disappears before you do any other work on the concrete. This may be an hour or two depending on temperature, humidity of the air, and the wind. During this time the concrete has begun to cure and harden slightly.

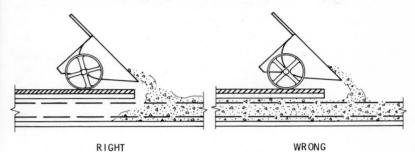

RIGHT   WRONG

Fig. 17. Showing the right and wrong way for pouring concrete. Pour farthest point first.

When pouring large areas of concrete, there is a right way and a wrong way. The right way is to provide the means for pouring the farthest point first, working back to the source of the concrete mix. This is illustrated in Fig. 17, showing boards laid across the forms

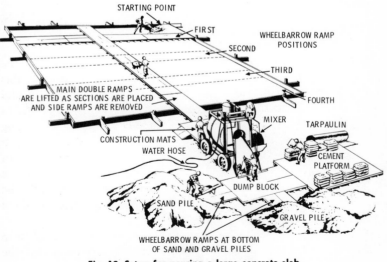

Fig. 18. Setup for pouring a large concrete slab.

or screeds. The boards are placed close enough together to make a runway for the wheelbarrow to take the concrete to the farthest point first. As the areas are filled, boards are removed for pouring the next load. Cross-board separators may be laid between the forms to make it possible to screed or level a section at a time. If a long area is poured at one time, the first concrete to be poured may begin to take a set and make leveling more difficult. As the strike-off board is drawn forward, excess concrete is allowed to fall into the next preceding section.

Fig. 18 shows the complete setup for pouring a concrete tennis court or the slab of a barn or house. Note how the material is set out for easy dumping into the mixer, in addition to the wheelbarrow runways for carrying the concrete to the farthest point first.

When pouring concrete from a *ready-mix* truck, omit the edging form at the entrance for the truck. Provide means for the chute of the truck to reach the last section first. *Ready-mix* trucks are equipped with extension chutes which may be added to the fixed chute on the truck. Chutes are on a swivel for laying the concrete exactly where needed. Since *ready-mix* trucks work fast, ample help must be on hand for leveling as the pouring takes place, or for wheelbarrows to haul to any position where truck chutes cannot reach.

CHAPTER 5

Reinforced Concrete

Concrete has tremendous compressive strength and durability. These are the reasons for its popularity in many construction projects. However, it has poor strength in tension. A beam made only of concrete and supported at its ends, would not bear much weight at its center. Fig. 1 is a sketch of a concrete beam with stresses at the points shown with arrows. Cracks would develop at the small hash marks, and the concrete would soon give way.

By adding steel supports embedded into the concrete, its tensile strength is improved considerably. Steel reinforcement can add the equivalent of six times added thickness of concrete for a given load. Fig. 2 shows a cross section of a concrete beam with reinforcing bars embedded near the bottom. The bars act in opposition to the stress at the bottom as a result of loads at the top.

Hollow concrete columns, frequently used to support upper floors and roofs, have a tendency to bulge outward with heavy top loading. This is illustrated in Fig. 3 for both square and round columns. Steel reinforcing supplies the lateral strength to overcome the effect of bulging.

Until concrete is fully cured, which may take up to a year, it has a tendency to creep due to plastic flow. If subjected to loads after the initial curing (typically 28 days), concrete may become slightly deformed. Steel reinforcement reduces creep considerably. After a period of time, perhaps one year, all plasticity in concrete is gone, and creeping no longer occurs.

87

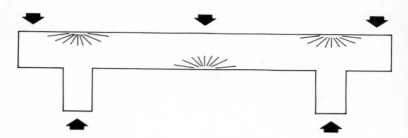

Fig. 1. A concrete beam showing stress points.

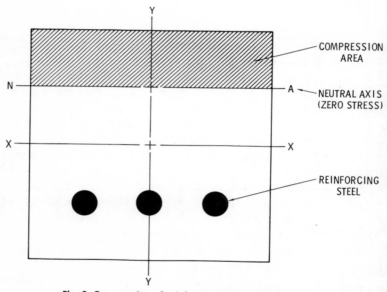

Fig. 2. Cross section of reinforcement in a concrete beam.

Of importance in the successful reinforcement of concrete by steel is good bonding between the steel and the concrete. Reinforcing steel must be free of rust, scale, oil, or any surface material that could affect bonding. If a good bond does not exist between the steel and the concrete, all the beneficial effects of reinforcement is lost. Plain-surfaced bars have been replaced by bars with a rough

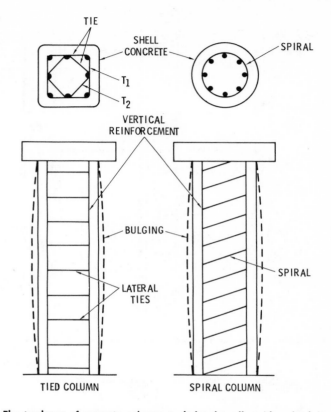

Fig. 3. The tendency of concrete columns to bulge laterally with a load is overcome by lateral steel reinforcement.

design on the outside, called *deformed* bars. Fig. 4 shows three examples of the surface texture on deformed bars. Concrete flows into and around the deformations and, when cured, provides a perfect bond to the steel.

REINFORCING RODS AND SCREEN

Reinforcing rods or bars may be round or square. Smaller sizes are usually round, while those over 1″ in diameter may be round or square. Table 1 shows the bar numbers and their sizes.

89

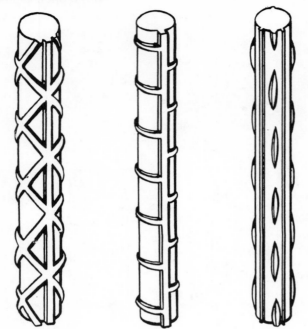

Fig. 4. Deformed bars are used for better bonding between concrete and steel bars.

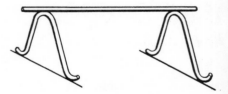

HIGH CHAIR -HC

CONTINUOUS HIGH CHAIR -CHC

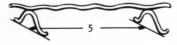

SLAB BOLSTER -SB

BEAM BOLSTER -BB

Fig. 5. Chairs or bolsters used to support bars in concrete beams.

Table 1. Reinforcing Bar Numbers and Dimensions

Bar sizes*		Weight (lbs. per. ft.)	Cross-sectional area (sq. in.)
Old (inches)	New numbers		
¼	2	0.166	0.05
⅜	3	0.376	0.1105
½	4	0.668	0.1963
⅝	5	1.043	0.3068
¾	6	1.502	0.4418
⅞	7	2.044	0.6013
1	8	2.670	0.7854
1/square	9	3.400	1.0000
1⅛ square	10	4.303	1.2656
1¼ square	11	5.313	1.5625

*Note. The new bar numbers are based on the nearest number of ⅛ inch included in the nominal diameter of the bar. Bars numbered 9, 10, and 11 are round bars and equivalent in weight and nominal cross-sectional area to the old type 1″, 1⅛″ and 1¼″-square bars.

To support bars the proper distance above the bottom of concrete beams or slabs, *chairs* or *bolsters* are used. Fig. 5 shows four common types. Depending on the application, bar assemblies may contain hooks, trusses, and stirrups, as well as bolsters welded into place. Fig. 6 shows an example and names the parts. To reduce handling on the job, reinforcing steel trusses may be made to order for a specific need. Fig. 7 is one example of this. This one is called the *Kahn* trussed bar, a popular configuration over the years, but

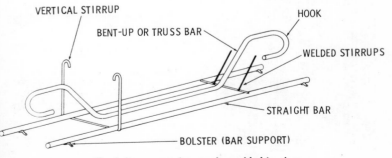

VERTICAL STIRRUP

BENT-UP OR TRUSS BAR

HOOK

WELDED STIRRUPS

STRAIGHT BAR

BOLSTER (BAR SUPPORT)

Fig. 6. Bar accessories may be welded in place.

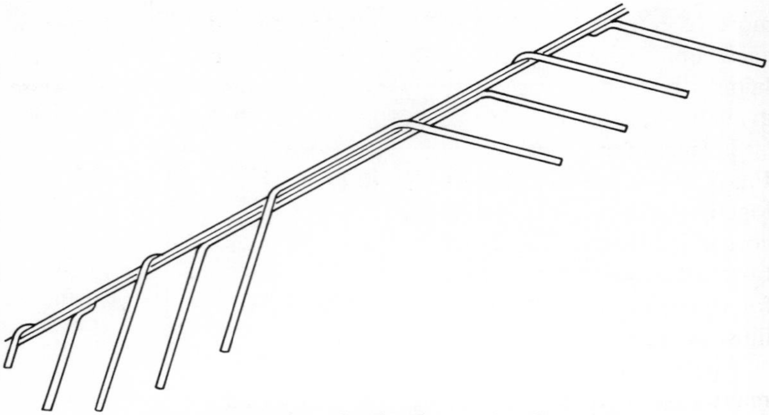

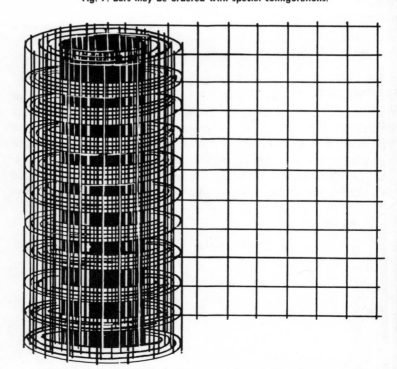

Fig. 7. Bars may be ordered with special configurations.

Fig. 8. Concrete slab reinforcement is usually in the form of a screen mesh.

more recently replaced by made-to-order methods which better fit each job. Small jobs use straight stock or hooked bars which are laid on bolsters in a crosswise pattern. For thin slabs, screens are available in rolls, as shown in Fig. 8.

Steel reinforcement in columns may take a number of forms. Fig. 9 shows three of them. In figure (A) are the vertical rods held together with transverse steel wire. Cross-bracing from corner to corner is often added. In figure (B) is a spiral coil of steel wire fastened to vertical rods for support. This is the most popular type for round columns. Heavy columns may include steel T-beams, as illustrated in figure (C).

Another type of slab reinforcing steel is the expanded steel sheet made in several patterns. Two of these are shown in Fig. 10. Expanded metal is made from flat steel stock and punched with a pattern. When pulled apart, they form an open mesh.

HOW TO USE REINFORCEMENT

On industrial buildings, dams, and other heavy concrete construction projects, the architectural and engineering drawings for the structure also include details for the placement of reinforcement members in the concrete. Unless otherwise permitted by the plans, rods must never be bent after they are placed in the concrete. A

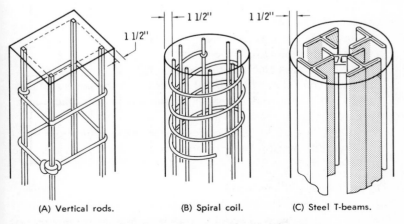

(A) Vertical rods.　　(B) Spiral coil.　　(C) Steel T-beams.

Fig. 9. Three forms of concrete column reinforcement.

93

few general rules which are followed by architects and engineers are listed below. These have to do with the thickness of the concrete adjacent to the reinforcement.

Fig. 10. Expanded-metal slab reinforcement.

Where concrete is deposited against the ground without the use of forms—not less than 3″ of concrete is used around the reinforcing steel bars.

Where concrete is poured on the ground, but in forms—not less than 2″ of concrete is used around the reinforcing steel bars.

In slabs and walls not exposed to ground or weather—not less than ¾″ of concrete is used. In beams, columns, and girders not exposed to the ground or weather—not less than ½″ of concrete is used.

In all cases—concrete thickness must be at least equal to the diameter of the bars.

Bars must be maintained clean of rust or other coatings to assure good bonding with the concrete.

Bars bent on the job must be cold bent.

Reinforcement must be placed according to the plans and supported with approved chairs, concrete blocks, or other metal spacers.

No splices shall be made unless the plans call for them.

All welding must follow the approved method outlined by the *American Welding Society* (AWS D12.1).

Reinforcement is applied near the area of greatest tension stress. This is usually near the surface opposite the load. This is shown in the cross-sectional view of Fig. 1. Large slabs poured on the ground,

RECOMMENDED SIZES - 180° HOOK	J	BAR EXTEN.	APPROX. H	BAR SIZE D
BAR EXTENSION REQUIRED FOR HOOK D = 6d FOR BARS #2 TO #7 D = 8d FOR BARS #8 TO #11	2 3 4 5 6 7 10 11 1/4 12 1/2 14	4 5 6 7 8 10 13 15 17 19	3 1/2 4 4 1/2 5 6 7 9 10 1/4 11 1/4 12 3/4	#2 3 4 5 6 7 8 9 10 11

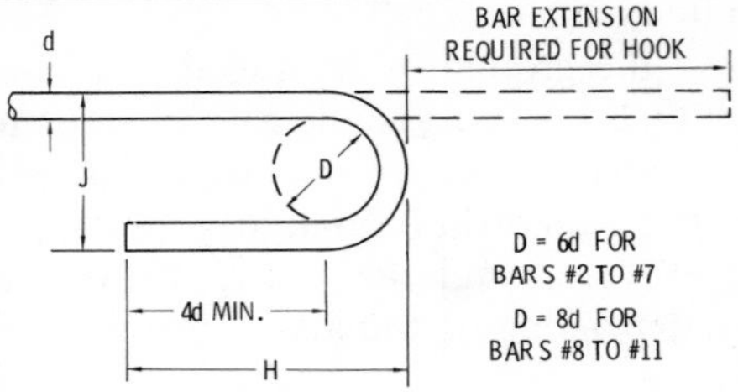

MIMIMUM SIZES - 180° HOOK	J	BAR EXTEN.	APPROX. H	BAR SIZE D
EXT. D = 5d MIN. D= 5d MAX. NOTE: MINIMUM SIZE HOOKS TO BE USED ONLY FOR SPECIAL CONDITIONS; DO NOT USE FOR HARD-GRADE STEEL.	1 3/4 2 3/4 3 1/2 4 1/4 5 1/4 6 7 8 9 10	4 5 5 6 7 9 10 11 13 14	3 1/2 4 4 1/4 4 3/4 5 3/4 6 1/2 7 1/2 8 1/2 9 1/2 10 1/2	#2 3 4 5 6 7 8 9 10 11

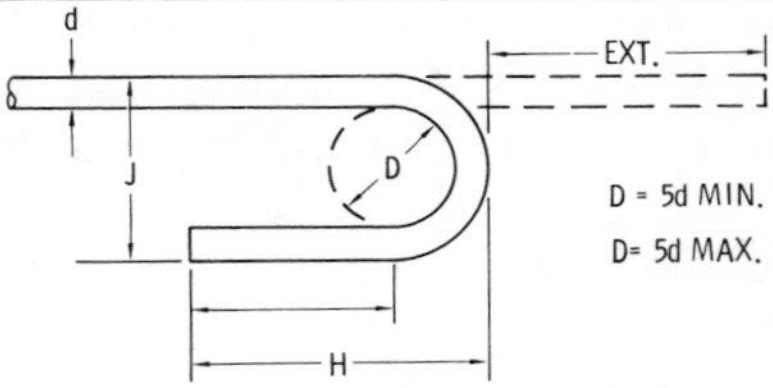

RECOMMENDED MINIMUM SIZES - 90° HOOK	BAR EXTEN.	APPROX. J	BAR SIZE D
EXT. 12d MIN. D = 7u	3 3 3 4 4 5 6 7 8 9	3 1/2 4 4 1/2 5 6 7 9 10 11 1/4 12 1/2	#2 3 4 5 6 7 8 9 10 11

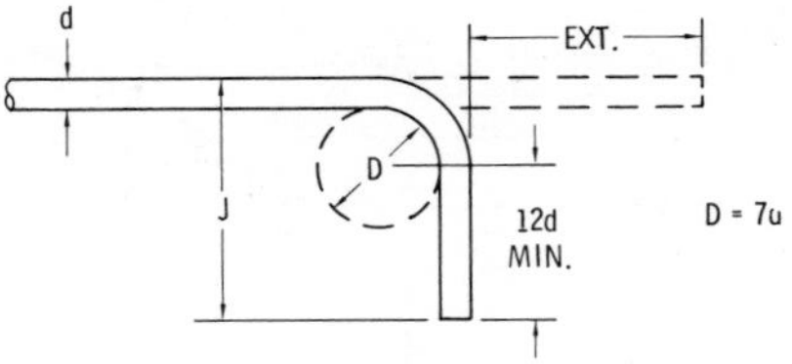

RECOMMENDED SIZES - 135° STIRRUP HOOK	BAR EXTEN.	H	BAR SIZE D
H EXT. D = 5d NOTE: STIRRUP HOOKS MAY BE BENT TO THE DIAMETER OF THE SUPPORTING BARS.	3 1/2 4 4 1/2 5	2 2 1/4 2 1/2 2 3/4	#2 3 4 5

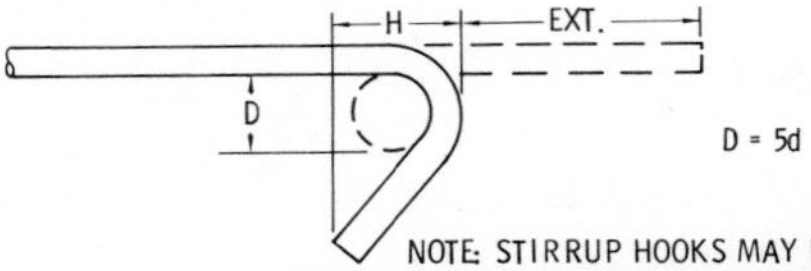

Fig. 11. Hook dimensions for on-the-job bending of reinforcing bars.

95

as for factory floors, barn floors, and house slabs (where there is no basement or crawl space) will be guided by architects drawings specifying the sizes and location of the reinforcing bars. The average distance from the ground is about 1½".

The bars are held above the ground by high chairs or bolsters, as previously described. Often it is acceptable to use other forms of raising the bars, such as precast concrete blocks, clean rocks of fairly uniform size, or anything that will become part of the poured concrete. Cross bars are also laid, and either wire-tied or welded to the perpendicular bars, depending on the mass of concrete to be poured.

On large jobs, bonding between the steel bars and the concrete must depend on the irregular surface of the bars alone. The ends of the bars must be bent to form hooks. The hooks make a firm grip on the cured concrete for greater strength and reduced shrinkage during hydration.

For large construction projects where a large number of reinforcing bars are ordered, their length may be specified and the hooks provided by the steel fabricator. On smaller jobs where stock

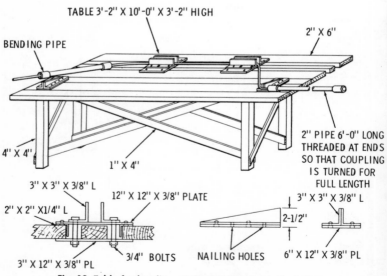

Fig. 12. Table for bending reinforcement bars on the job site.

bars are used, it becomes necessary to bend the hooks on the job. Hooks must be cold bent. Fig. 11 shows the recommended radius of the bend and the length of stock to provide at the end. Fig. 12 shows the construction of a table for bending reinforcing bars. In emergencies, a hickey may be made from a 2″ × 1½″ pipe tee with a 3-ft. length of 1½″ pipe screwed into the center junction. Saw off one arm of the tee to make sharper bends.

Smaller slabs are reinforced with welded wire screen (available in rolls) as shown in Fig. 8. Pieces are cut off to size with a hack-

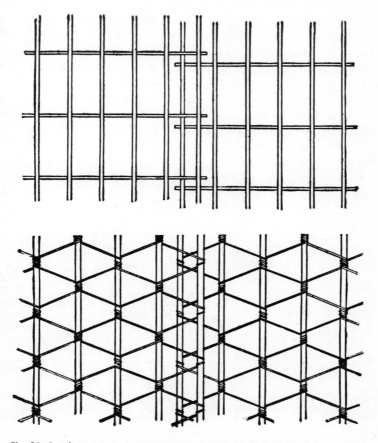

Fig. 13. Overlap wire mesh reinforcement where greater width is to be covered.

saw or heavy-duty wire cutters. Small-size bolsters are used to support the material above the ground. Small rocks may also be used if they are first washed clean. For a section of 4″ thick driveway or walk, for example, the mesh only needs to be about 1″ above the ground. Another method frequently used is to lay the screen on the ground, pour the concrete, then reach a hook down into the concrete and lift the screen up about 1″. Remove the hook, then spade and level the concrete. The wire mesh will stay in the position to which it was lifted.

Where the width of the mesh roll is not wide enough to reach to the edges of the concrete, an overlap of screening is important to good strength. This is shown for two forms of screening in Fig. 13. The illustration in Fig. 14 shows roll-type mesh for a concrete slab to be laid for a home. This is a common type of home construction in the southwest. Wires are welded where they cross.

The need for reinforcement in footings will depend on the load to be supported. For example, footings for block walls around homes are sometimes poured into a ditch without reinforcement.

Fig. 14. Roll-type mesh reinforcement for a concrete slab.

PRECAST CONCRETE BLOCK
OR STONE OF PROPER SIZE

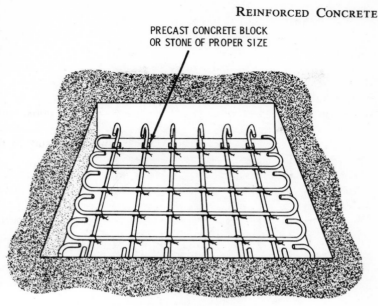

Fig. 15. Reinforcement of a footing for a column.

While the concrete base may be capable of supporting a 5-ft. high block wall, it could easily develop cracks due to subsoil settling. This settlement will cause cracks at the mortar lines in the block wall. Steel reinforcement rods should be laid in the ditch or form (if used), and supported just above the ground level. High and thick industrial walls, whether of concrete block or poured concrete, must be built on footings with steel reinforcement. Footings for columns, load-bearing walls, and beams must be reinforced.

Fig. 15 shows the bars laid in a typical footing for a column. The reinforcing steel bars are laid in place after the forms are in place. Crossed bars are tied together with wire and held above the subgrade by any of the means mentioned before.

Fig. 16 shows a method of installing reinforcement in a concrete poured wall. Individual bars, tied where they cross, or welded wire fabric may be used. A convenient method for holding the material in place, centered in the concrete, is to hang them from wood blocks fastened to one side of the form. As the concrete is poured and nears the top of the form, remove the wood blocks. Then complete the pouring to the top.

99

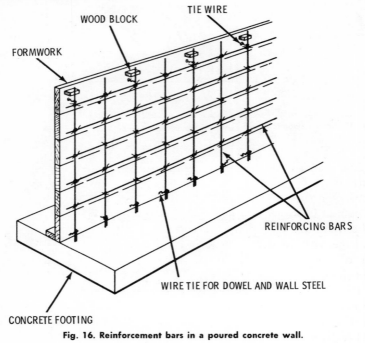

Fig. 16. Reinforcement bars in a poured concrete wall.

Concrete beams must have steel reinforcement. The placing of the bars depends on the direction of stress, which in turn depends on the purpose of the beam. Fig. 17 is an example of a beam supported on a column, and on which the principal stress is upward in the center and downward at the ends.

PRESTRESSED CONCRETE

In long beams, as for use on bridges, increased strength and a lower volume of concrete and steel can be obtained by prestressing. Steel used in prestressed concrete is a special high-strength type capable of withstanding a pull of nearly 200,000 psi without excessive stretch. The steel rods used pull in on the ends of the beam and place the concrete under high compression to overcome the effect of any bend on the part of the concrete and to eliminate the effects of tensile pressures.

100

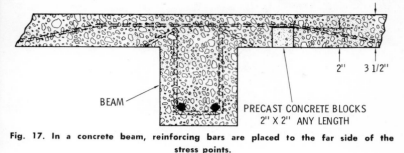

2"   3 1/2"

BEAM

PRECAST CONCRETE BLOCKS
2" X 2" ANY LENGTH

Fig. 17. In a concrete beam, reinforcing bars are placed to the far side of the stress points.

Two methods of prestressing are used—pretensioning and post-tensioning. In pretensioning, high-strength steel rods are stretched between abutments on a casting bed, then stretched. The concrete is poured in forms through which the rods (under tension) pass. After thorough curing, the rods are released from their abutments, and the cast concrete with the rods under tension inside is lifted out of place. Fig. 18 illustrates this. The bond between the rods and concrete holds the concrete under tension.

Post-tensioning consists of pouring concrete in a form with rods or bars, not under tension, passing through. The rods are treated so there is no bond between it and the concrete. One end of the rod is fixed, with a heavy washer fastened in place. The other end has a heavy washer on it and is threaded for a large nut. After the concrete is thoroughly cured, pneumatic equipment connected to the free end of the rod will turn up on the nut, or other tightening device, thus putting the concrete under tension. The rod must be free to move within the concrete for post-tensioning.

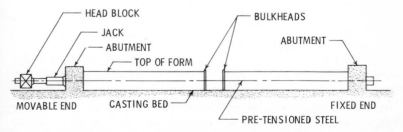

HEAD BLOCK   BULKHEADS

JACK   ABUTMENT
ABUTMENT

TOP OF FORM

MOVABLE END   CASTING BED   FIXED END
PRE-TENSIONED STEEL

Fig. 18. In pretensioned, prestressed steel reinforcement, the concrete is poured around rods under tension.

101

CHAPTER 6

Concrete Forms

Since a concrete mixture is semifluid, it will take the shape of anything into which it is poured. Accordingly, molds or forms are necessary to hold the concrete to the required shape until it hardens.

Form work may represent as much as one-third of the total cost of a concrete structure, so the importance of the design and construction of this phase of a project cannot be overemphasized. The character of the structure, availability of equipment and form materials, anticipated repeated use of the forms, and familiarity with methods of construction influence design and planning of the form work. Forms must be designed with a knowledge of the strength of the materials and the loads to be carried. The ultimate shape, dimensions, and surface finish must also be considered in the preliminary planning phase.

NEED FOR STRENGTH

Forms for concrete structures must be tight, rigid, and strong. If forms are not tight, there will be a loss of concrete which may result in honeycombing, or a loss of water that causes sand streaking. The forms must be braced well enough to stay in alignment and strong enough to hold the concrete. Special care should be taken in bracing and tying down forms such as those for retaining walls in which the mass of concrete is large at the bottom and tapers toward the top. In this type of construction, and in other types such as the

Fig. 1. Steel forms in place for a concrete slab.

first pour for walls and columns, the concrete tends to lift the form above its proper elevation. If the forms are to be used again, they must be easily removed and re-erected without damage. Most forms are made of wood, but steel forms are commonly used for work involving large unbroken surfaces such as retaining walls, tunnels, pavements, curbs, and sidewalks, as shown in Fig. 1. Steel forms for sidewalks, curbs, and pavements are especially advantageous since they can be used many times.

WOOD FORMS

The majority of concrete building construction is done by using wooden forms. Various woods are used, the selection depending on the character of the work and the available supply in the local lumber yards. In general, forms are best made of any of the soft woods since they do not warp as easily when wet.

Contrary to the usual practice in building construction, green lumber or lumber which is only partly air dried will keep its shape in concrete forms for rectangular construction better than lumber that is kiln dried. If kiln dried lumber is used, it should be thoroughly wet before the concrete is poured. This is to keep the lumber from absorbing water from the concrete; if the forms are made tight (as they should be) the swelling from absorption will cause the forms to buckle or warp. Oiling or greasing the inside of the forms before use is also recommended, especially where they are to be reused. Oiling prevents absorption of water, assists in keeping them in shape when not in use, and makes their removal much simpler.

Spruce seems to be the best all-around material. It can readily be obtained in almost any locality and is undoubtedly an excellent lumber to use for joists, studs and posts. For sheathing, however, white pine is better than spruce because of its smoothness and its resistance to warping, but this wood is generally too expensive (except for cornice and ornamental work) and spruce makes a good substitute. If white pine is used for sheathing, it should be noted that this kind of lumber is not durable because it is very soft and is accordingly not desirable where the forms are to be used many times. Norway pine and Southern pine are generally the most available and give satisfaction.

Hemlock is not usually desirable, especially for that part of form work which comes into contact with the concrete, but sometimes it is used for ledges, studs, and posts. This wood is too coarse-grained to be suitable for sheathing and is liable to curl when exposed to the weather or to wet concrete. Form lumber should be free from loose knots or other defects and irregularities that would be reproduced by the concrete. It is often possible to build forms from stock lengths of lumber without cutting, thereby saving material and labor. It is seldom economical to rework second-hand lumber for forms.

All lumber should be dressed at least on one side and both edges. Since forms are cleated, dressing is necessary in order that the face next to the concrete will be uniform. In footings and rough work that is not to show, practically any lumber can be used that will hold wet concrete, but for forms that are to be used again, the additional care of cleaning will pay the cost of using smooth lumber.

In face work where a smooth and true surface is important, the lumber employed should be dressed on all four sides. The edges may be cut square, mitered, or tongue-and-groove. The latter joint is nearer watertight and tends to prevent warping. Beveled-edge joints are desirable where dry lumber is used since buckling caused by the edges crushing as the boards swell is prevented. The tongue-and-groove joint, however, gives the best results under ordinary conditions. The tongue-and-groove joint is more expensive than the beveled-edge joint, but it gives smoother surfaces with repeated use.

Any concrete laid below ground level for support purposes, such as foundations, must start below the freeze line. This will vary for different parts of the country, but is generally about 18″ below ground level. The length of time necessary to leave the forms in place depends on the nature of the structure. For small construction work where the concrete bears external weight, the forms may be removed as soon as the concrete will bear its own weight; that is, between 12 and 48 hours after the concrete has been poured. Where the concrete must resist the pressure of the earth or water (as in retaining walls or dams), the forms should be left in place until the concrete has developed nearly its final strength; this may be as long as three or four weeks if the weather is cold, or if anything else prevents quick curing.

Forms must be watertight, rigid, and strong enough to sustain the weight of the concrete. They must also be simple and economical, and if they are to be used again, designed so that they may be easily removed and re-erected without damage to themselves or to the concrete. The different shapes into which concrete is formed means that each job will present some new problems to be solved, but there are typical forms that will cover a large part of concrete construction.

Since concrete weighs from 130 lbs. to 150 lbs. per cu. ft., it must be evident that forms should be of substantial construction and properly proportioned so as to be rigid against pressures due to the weight of material, thus avoiding any bulging of vertical or sagging of horizontal forms. In the horizontal forms, any sagging will result in small cracks developing on the surface while the concrete is hardening. These cracks will gradually widen, which will prevent the construction from having the desired strength.

To retain the concrete in proper position in its plastic condition until it hardens, the forms should be constructed of:

1. Retaining boards.
2. Supporters, or studs.
3. Braces.

The ordinary arrangement of these members is shown in Fig. 2. The thickness of the lumber varies according to its use. For short spans between supports, such as floor slabs and wall forms, 1″ stock is generally used. For columns, either 1″ or 1¼″ lumber is

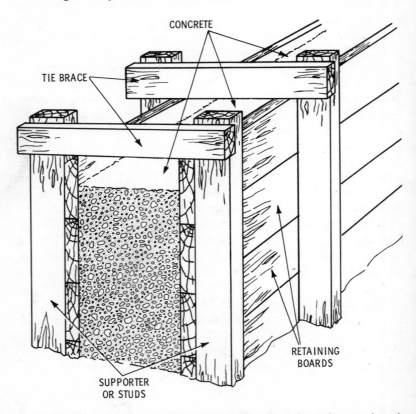

Fig. 2. Elements of a concrete form showing the principal members used, such as the retaining boards, supports, and braces.

used, according to the spacing of the yokes. For beam sides and bottoms, 2″ material is used. For supporters, 2″ × 4″, 2″ × 6″, or in extreme cases, 2″ × 8″ studs are used. The volume of the concrete and its depth determines the dimension and spacing of the support. Forms constructed of 2″ × 4″ supporters with 1″ retaining boards should have studs spaced not more than 2 ft. apart. This is to prevent bulging of the sheathing when subjected to the pressure of the concrete until it hardens.

TYPES OF FORMS

Depending on application, forms may be very simple or very complex. The simpler the form, the easier the job, and the more economical. Bear in mind one thing—the form must be strong enough to support the concrete rigidly for a day or two until the concrete has taken a firm set.

The simplest form is one used for small column supports, like that of Fig. 3. An illustration of its use is shown in Fig. 4, in which it holds the concrete footing for a 4″ × 4″ post for a patio cover. If the form is made of redwood, it may be left in place as shown in the illustration. The earth itself may be the form if the sides are dug straight and smooth. Fig. 5 shows a trough dug as a footing

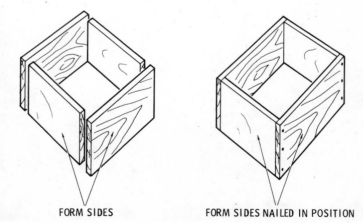

FORM SIDES   FORM SIDES NAILED IN POSITION

Fig. 3. A simple box-like structure which will serve for small columns and posts.

for a concrete block wall. Long troughs are dug with a power shovel, but the sides must be finally smoothed by hand.

Another example of the earth as a form is a simple clothesline pole installation. A pole digger will cut about a 6″ diameter hole. Set the pole in place in the hole, and pour concrete into the hole up to about 2″ from the top of the soil. Cover the remainder with soil. The pole must be left undisturbed for several days for curing before use.

Slab concrete only requires simple 2″ × 4″ or 2″ × 6″ boards, as described in Chapter 4, or metal rails as shown in Fig. 1. Since the pressure of a few inches of concrete is not great, the form boards may be staked to the earth.

Fig. 4. The form is being used to secure a 4″ x 4″ post in concrete.

Fig. 5. A concrete block wall footing can be poured in the earth if the ditch is dug straight and smooth.

FOOTINGS AND FOUNDATIONS

There is no substitute for an adequate foundation, a key part of every building. A concrete foundation represents only a small part of a typical building's cost and is one of the best investments in construction. An adequate footing provides a stable base and directly affects both the life and performance of the building. In addition, it gives protection against rats, mice, termites, water, and the elements.

A foundation consists of:

1. Its *bed*—the earth giving support.
2. Its *footing*—the widened part of the structure resting upon the bed.
3. Its *wall*—the structural part resting upon the footing.

The size of the footing depends on the load-carrying capacity of the soil and the weight of the building and its contents. Soils vary in their ability to support weight. The load-carrying capacity of several common soils is given in Table 1. Firm clay, for example, has a carrying capacity of 2 tons per sq. ft., meaning that under normal conditions, 1 sq. ft. of firm clay will support 4000 lbs.

Table 1. Load-Carrying Capacities of Soils

Type of soil	Tons per sq. ft.
Soft clay	1
Firm clay or fine sand	2
Compact fine or loose coarse sand	3
Loose gravel or compact coarse sand	4
Compact sand-gravel mixture	6

Courtesy Portland Cement Association.

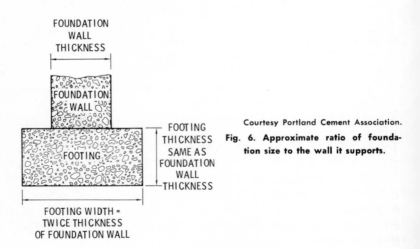

Courtesy Portland Cement Association.

Fig. 6. Approximate ratio of foundation size to the wall it supports.

A common rule of thumb often used to dimension footings for lightly loaded buildings is that the footing shall be twice as wide as the foundation wall and as thick as the wall is wide. See Fig. 6. By this rule, the following footing sizes would be used:

For 8″ thick foundation wall (or less), 8″ × 16″ footing.
For 10″ thick foundation wall, 10″ × 20″ footing.
For 12″ thick foundation wall, 12″ × 24″ footing.

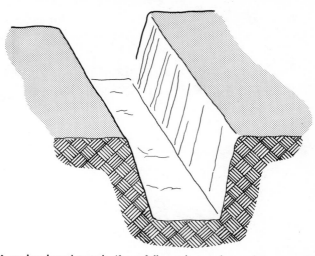

Fig. 7. A wedge-shaped trough, if carefully made, can be used to support the side boards of a footing form.

If the structure is heavily loaded, or the soil conditions are questionable, consult an engineer or architect. In areas subjected to freezing, the bottom of the footing must be placed below the frost line to prevent frost-heave and resultant damage to the building. The trench bottom should be level and cut flat so the footing will bear evenly on undisturbed earth.

A wedge-shaped trough may be dug, like that shown in Fig. 7, in which case the sides of the trough may be used to support the form boards. This method and that of self-supporting side boards are shown in a cross-sectional view in Fig. 8. Fig. 9 shows a simple form and lumber dimensions for a footing for an above-grade wall. Fig. 10 shows a completed concrete foundation using the simple forms in Fig. 9. Where footings are well below ground, as for a full-basement house, drainage must be provided.

For wet soil conditions, and in order to assure proper drainage, a drain tile should be laid around the footing as shown in Fig. 11.

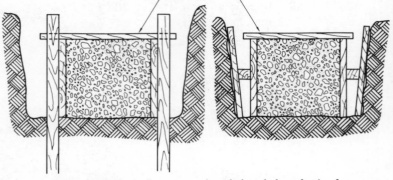

Fig. 8. Two methods used to support the side boards for a footing form.

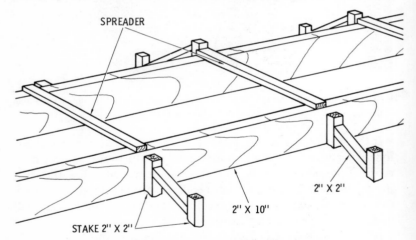

Fig. 9. Simple form for a typical concrete foundation.

The drain tile should be laid with open joints and drained to a suitable outlet with a slope of 1″ every 12 ft. In no case should the tile be lower than the footing. Joints between the tile should be covered with strips of tar paper, or roofing felt, to prevent sediment filling the tile during back filling. The tile line should be covered with not less than 18″ of gravel or cinder fill. Filling of dirt around the footing should be delayed until after the sub-floor is completed.

Fig. 10. The complete house foundation using simple forms shown in Fig. 9.

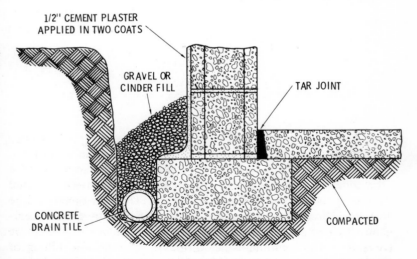

Fig. 11. Drain tile should be included where foundations are made for basement wall support.

113

WALL FORMS

Wall forms represent construction of greater complexity. Because of the weight of concrete, forms must be strong enough to hold their shape. Certain minimum board size and supporting members should be observed.

The following definitions apply to the members of the form identified in Fig. 12.

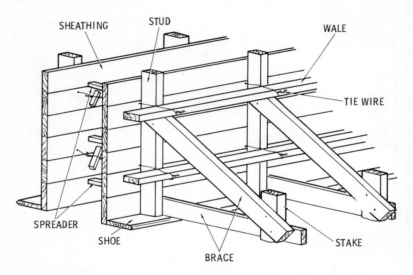

SHEATHING   STUD   WALE   TIE WIRE   SPREADER   SHOE   BRACE   STAKE

Fig. 12. Basic wall forms and the supporting members used to hold the concrete forms securely.

Sheathing—Sheathing forms the inner surface of the box-like structure. If the concrete is to be exposed, this surface must be as smooth as possible to reduce the need for finish treatment of the concrete. Otherwise, the surface must be at least free from cracks or holes, and must be watertight. The best material for a tight fit is tongue-and-groove boards. Outdoor-type plywood, and sometimes *Masonite,* can be used.

Studs—The sheathing is held firmly in place against the pressure of the concrete by vertical studs. These are usually 2″ × 4″ lumber, but high thick walls require larger size material.

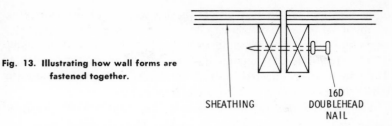

Fig. 13. Illustrating how wall forms are fastened together.

SHEATHING

16D
DOUBLEHEAD
NAIL

Wales—High walls require reinforcement of the studs. This is done with horizontal bracing, called wales. They may be the same material as the studs.

Braces—To shore up the members, braces are added and staked to form approximately a 30° angle to the vertical. Braces may take many forms and are a matter of judgment as to how shoring is needed to resist lateral loads, including the effects of the equipment used to pour the concrete.

Depending on the length of the wall, forms may be made as a single unit or in sections. Remember, unless the forms are being made for a single job never to be duplicated, the forms must be

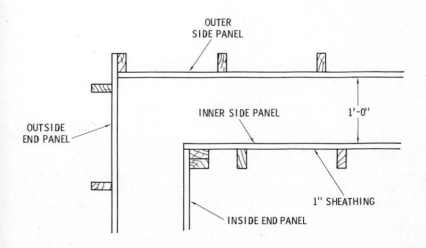

Fig. 14. Details for fastening walls together at corners.

115

removable and reusable. Wall panels should be made up in lengths of about 10 ft. so that they can be easily handled. The panels are made by nailing the sheathing to the studs. Sheathing is normally 1″ (13/16″ dressed) tongue-and-groove lumber, or ¾″ plywood. The panels are connected together as shown in Fig. 13. Form details at the corner of a wall are given in Fig. 14.

The sides of the wall form must be kept apart with spreaders

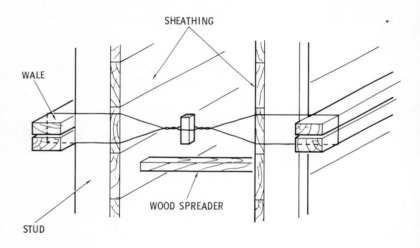

SHEATHING

WALE

WOOD SPREADER

STUD

Fig. 15. Wire ties used to hold form walls together which prevents bulging.

until the concrete is placed. They must also be tied together so they will not spread apart as the concrete is poured. Two methods of doing this with wire ties are shown in Figs. 15 and 16. The method shown in Fig. 15 illustrates the use of wire ties; this method should be used only for low walls and when tie rods are not available. The wire should be No. 8 or No. 9 gauge soft black annealed iron wire, but in an emergency barbed wire can be used. The individual ties should have the same spacing as the stud spacing, but never more than 3 ft. A spreader should be placed near each tie wire. Each tie is formed by looping a wire around the wale on one form side. It is then brought through the forms, crossed inside the forms, and looped around the wale on the opposite side. The

116

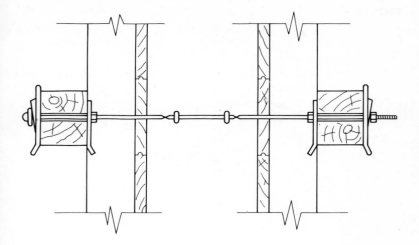

Fig. 16. Bolts are commonly used to serve as ties and spreaders.

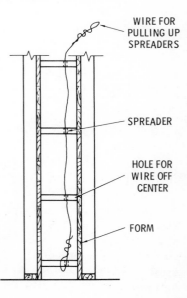

WIRE FOR PULLING UP SPREADERS

SPREADER

HOLE FOR WIRE OFF CENTER

FORM

Fig. 17. Suggested way to remove spreaders as the concrete is poured.

117

CONCRETE FORMS

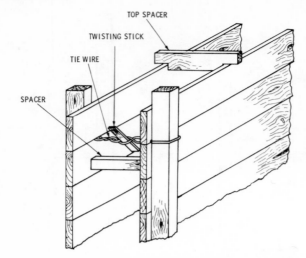

(A) Twisted wire ties are used with spacers to hold the concrete form together.

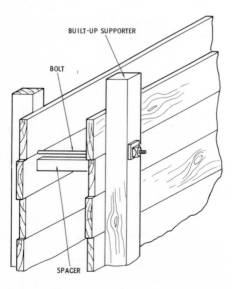

(B) Bolts are used, together with spacers, to hold a concrete form.

Fig. 18. Four suggested methods of

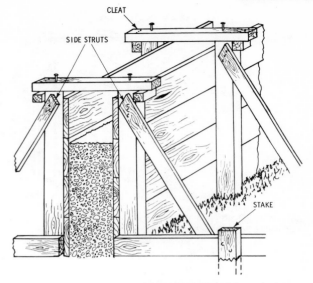

(C) Continuous wall forms with side struts and through ties.

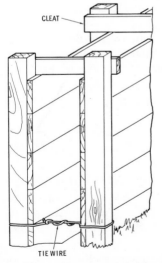

(D) Vertical supports are held in place using twisted wire and cleats nailed in place.

constructing various concrete forms.

119

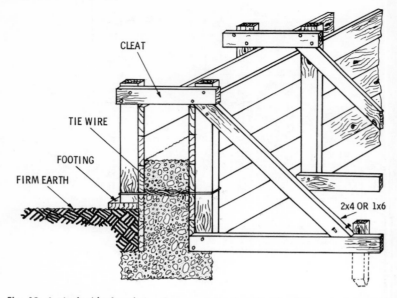

CLEAT

TIE WIRE

FOOTING

FIRM EARTH

2x4 OR 1x6

Fig. 19. A single-side foundation form used for forming foundations in firm earth.

wire is finally made taut by twisting. When the wedge method is used, the wire is first twisted around the point of the wedge. After this, the nail is removed and the wedge driven tight in order to draw the wire taut. The wood spreaders must be removed as the forms are filled so that they will not become embedded in the concrete.

A convenient way to remove these spreaders is shown in Fig. 17. A wire fastened to the bottom spreader is passed through holes drilled to one side of the center of each spreader. Pulling on the wire will remove the spreaders one after another as the concrete level rises in the forms. The tie rod and spreader illustrated in Fig. 16 serves both as a tie and a spreader. It should have the same spacing as tie wires. After the form is stripped, the rod is broken off at the notch which is slightly inside the concrete surface. If rust stains are not objectionable, and if the wall forms are to be maintained in exact position, this type of tie and spreader should be used. If appearance is important, the holes made by breaking off the tie bar should be grouted in with a mortar mix.

120

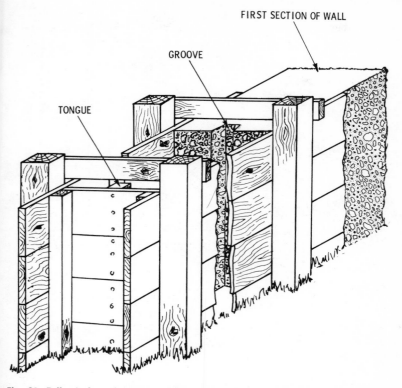

FIRST SECTION OF WALL

GROOVE

TONGUE

Fig. 20. Full-unit form for casting long walls in sections. The tongue at the closed end of the forms make a good bonding groove for newly deposited concrete.

Wall forms need not follow rigid rules of construction as long as they are solid and will take the pressure of the poured concrete. Fig. 18 shows four other styles of construction, any of which could serve the purpose.

Fig. 19 shows the method used for walls installed below grade. The earth is shown used as part of the support for two levels of grade. On long walls it is sometimes advisable to pour them in sections not solidly connected to each other. This permits some shifting of wall sections without cracks resulting, as in the case of sectionalized walks and driveways. Fig. 20 shows a wood tongue fastened to a lateral section of the form to make a groove at the

121

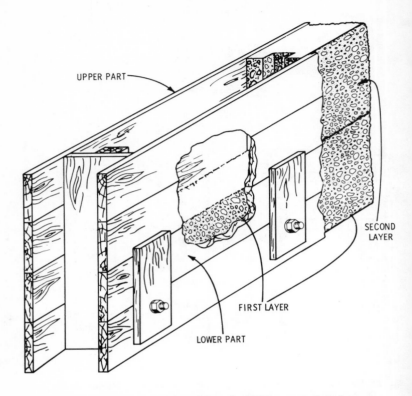

UPPER PART

SECOND
LAYER

FIRST LAYER

LOWER PART

Fig. 21. Layer-unit (sectional-type) form for building up walls in layers.

end of one section of wall. Obviously, the mating wall end must be made with a form of opposite contour.

On jobs where economy of lumber is important, it is possible to pour the concrete in sections, layer upon layer. This permits constructing forms only for each layer as the curing progresses. Fig. 21 shows a cutaway view of the layered sections. Fig. 22 shows how the same lumber may be used for each layer poured.

Openings for windows will require inserting a box-like structure between the form walls and securing in place. Fig. 23 shows this type of construction. Window frames are available in ready-made forms for placement in location. Be sure the concrete is well-

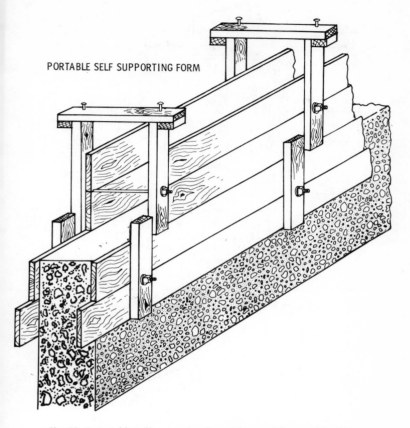

PORTABLE SELF SUPPORTING FORM

Fig. 22. A portable self-supporting form using a minimum of lumber.

spaded under the window frame before proceeding with pouring the balance of the wall.

COLUMN FOOTINGS

Wherever possible, the earth itself should become the form for a column footing. If a rectangular hole can be dug with vertical sides, concrete may be poured directly into it, with the necessary hardware for securing the column in place.

123

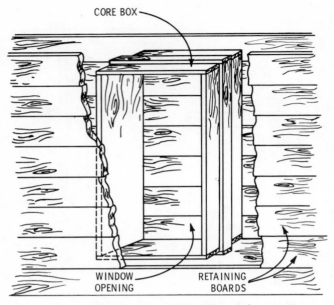

CORE BOX

WINDOW OPENING

RETAINING BOARDS

Fig. 23. Core box placed in a wall form for window opening.

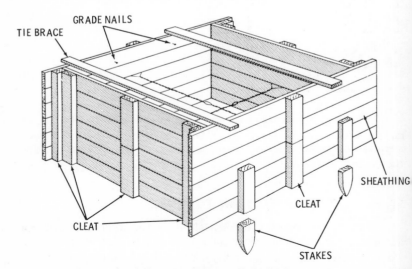

GRADE NAILS

TIE BRACE

SHEATHING

CLEAT

CLEAT

STAKES

Fig. 24. Basic form for a column footing.

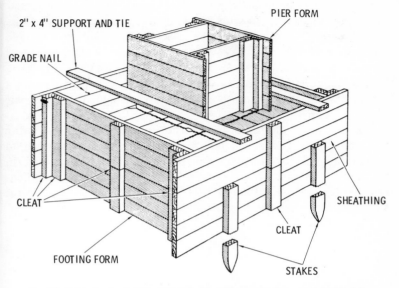

Fig. 25. Using the basic form shown in Fig. 24, a small pier may be added for support of a steel column.

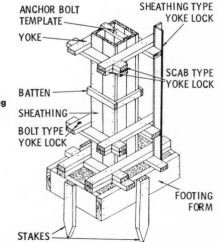

Fig. 26. Another combination footing and pier for lighter loads.

125

Footings may be made with wood structures in a manner similar to footings for walls, but with the shape and size for columns. Fig. 24 is a typical footing structure. The panel members are held together with No. 8 or No. 9 gauge iron wire wrapped around the center cleats as shown in the illustration. Note the difference in cleat handling for each pair of facing panels. This permits a tight corner fit, and allows the use of stakes and tie braces on one pair of facing panels.

Fig. 25 shows the same basic form, but for use with a footing and short pier, as a single pouring job. The pier is intended to support a steel column. This will require setting proper bolts in the center for tying to the column. Fig. 26 shows another combination footing and pier column. This footing and pier column is for steel columns supporting lighter loads.

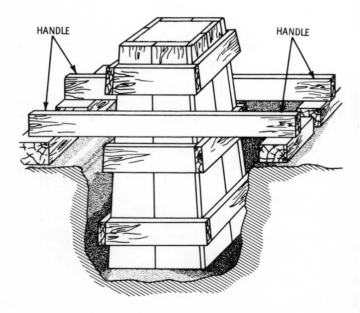

Fig. 27. Continuous form for a rectangular short pier or column. The pier can be leveled by putting blocks under the handles of the form.

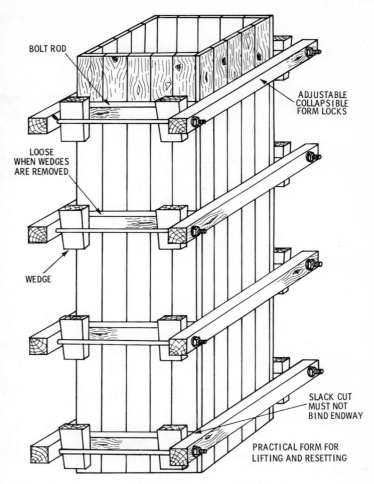

Fig. 28. Sectional forms for long rectangular columns, using through bolt ties and wedge adjustments.

COLUMN FORMS

Forms for pouring concrete columns are boxlike in construction, and are generally of two types:

1. Continuous.
2. Sectional.

127

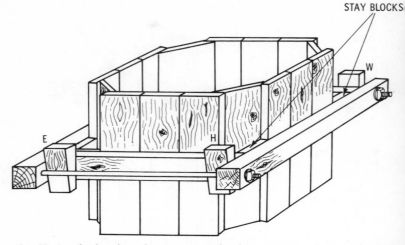

Fig. 29. Detail of a form for an octagonal column. Positions E, H, and W are wedges.

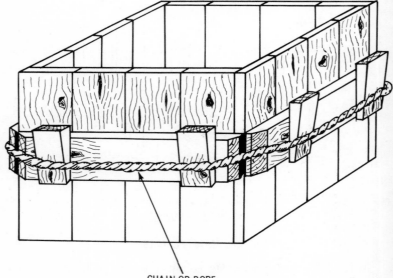

Fig. 30. Detail of a form for a rectangular column showing a method of clamping by means of a chain or rope using wedges.

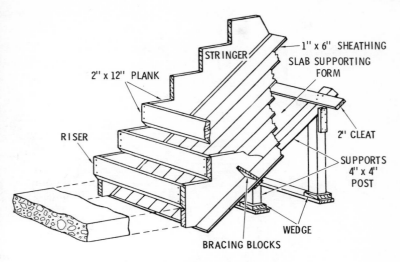

Fig. 31. A stairway form with essential supports shown.

The continuous form extends from the bottom to the top of the pier or column and is nonadjustable, being suitable for a short pier or column. In constructing such forms, it should be slightly tapered so that it can be lifted off after the concrete sets. See Fig. 27.

Larger columns, because of the great pressure near the bottom due to the weight of the concrete, must have more substantial forms, especially if cast in one pouring with continuous-type forms. Very heavy forms may be avoided by building up the column in several sections, using *sectional forms.* The forms are used several times for the several sections, and are held together by bolts and side pieces held in position by wedges as shown in Fig. 28.

After pouring one section, the form can usually be taken off within one or two days, and the form reset for the next section. For tapered columns, the reduction in the size of the column is obtained by removing a strip along one edge of the sides which do not have the yokes projecting and by removing a strip from one of the boards on the sides having projecting yokes. New bolt holes are either bored in the yokes, or packing strips are placed on the yokes for the wedges to bear against. If there is a boring

129

CONCRETE FORMS

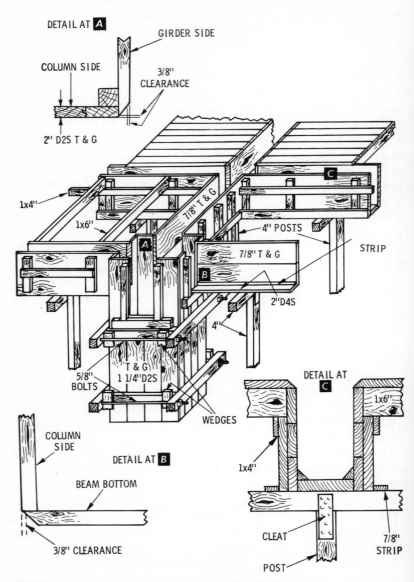

DETAIL AT **A**

GIRDER SIDE

COLUMN SIDE

3/8" CLEARANCE

2" D2S T & G

1x4"

1x6"

7/8" T & G

C

4" POSTS

STRIP

7/8" T & G

A

B

2" D4S

4"

T & G
1 1/4" D2S

5/8" BOLTS

WEDGES

DETAIL AT **C**

1x6"

1x4"

COLUMN SIDE

DETAIL AT **B**

BEAM BOTTOM

3/8" CLEARANCE

CLEAT

POST

7/8" STRIP

Fig. 32. Girder and beam construction showing assembled forms. Note A, B, and C exploded views.

machine on the job, the yokes should have a series of holes bored in them before they are fastened to the column sides so that it will not be necessary for the carpenters to bore them by hand.

Forms for octagonal columns are made as shown in Fig. 29. This column is identical with the one shown in Fig. 28, except that pieces are inserted in the corners to give the column eight sides. The flare is made at the top of the column by fitting in triangular pieces of wood. Since fresh concrete is practically liquid and over twice as heavy as water, the column form must be designed to withstand the bursting pressure of the concrete. This will make it necessary to have the yokes closer together at the bottom than at the top.

In place of bolts, rods with a malleable-iron clamp fastened in place with a set screw are often used for form work. These clamps may be obtained from supply houses. The clamp is slipped over the rod and brought to a firm bearing by a device furnished by the makers of the clamps. Then the set screw in the clamp is tightened, holding the clamp in place. A method of clamping columns by using a chain or wire rope is shown in Fig. 30. The chain is hooked around the column as tight as possible and the slack taken up with the wedges.

COMPLEX FORMS

Because of the plasticity of concrete, it may be poured into any shape for any application in construction projects. The construction of complex forms is generally specified in detail by engineering drawings. To illustrate the point, here are a few examples:

Forms for stairs require very careful construction. Because the concrete must carry its own load and that of the people using it, much of the pressure is tensile rather than compression. Unless the concrete is also supported by steel or concrete columns, it must include a steel reinforcement placed in the concrete, and the form must be extra rigid and left in place until complete curing has been achieved. Fig. 31 is an example of a simple stairway form. Note the use of 4″ × 4″ post supporting the form and the weight of the concrete to be poured.

Many buildings of concrete include concrete floors. Because of

131

CONCRETE FORMS

their weight, they must be supported on beams and girders, which
are in turn supported on columns. Fig. 32 is an example of one
such form structure. The construction of the form usually begins
with the separate construction of the slab (or floor) supports, the
beam, and the girder forms, which are built into the final forms,
including the column forms, as shown in the illustration.

FORM STRENGTH DESIGN

Concrete poured into forms develops hydrostatic pressures
which are developed by a number of factors. Ingredients of concrete are heavy. Their weight, the temperature of the concrete,
and the rate of placement all affect the hydrostatic pressure.

Standard formulas that take these factors into account have
been developed into graphs to aid in calculating the thickness of
sheathing, the size of studs, and column construction from pressures developed. Hydrostatic pressure is a function of the temperature of the plastic concrete and the rate of placement.

Table 2. Maximum Concrete Pressure Developed From Rate of Placing

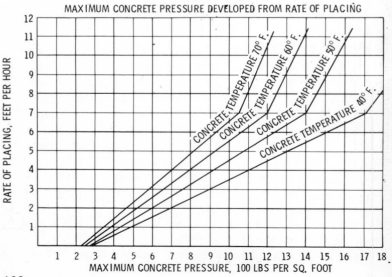

The chart in Table 2 shows the pressure that will be developed, knowing the temperature and rate of placement. Temperature can be measured with a thermometer, and the rate of placement is ft. per hour of height. For example, if concrete whose temperature is 70°F is poured for a 6-ft. high wall, and it is calculated that this much concrete can be poured in one hour, the pressure developed will be about 940 lbs. per sq. ft. This information is then used in the next two charts for form construction.

The chart in Table 3 will give sheathing thickness and stud spacing for pressures determined from Table 2. Stud spacing needs will vary with sheathing thickness. Carrying forward the same example, 1″ sheathing would require studs spaced every 14″ to hold a pressure of 940 lbs. However, 1¼″ sheathing would only require studs spaced every 19″ to provide the same strength. The chart in Table 4 sets the wale spacing for different stud sizes and sheathing thicknesses.

Pressure in a column is greatest at the bottom and decreases as you go up the column. It is greatest, of course, at the bottom of the highest column. Fig. 33 shows a typical column form construc-

Table 3. Sheathing Thickness and Stud Spacing

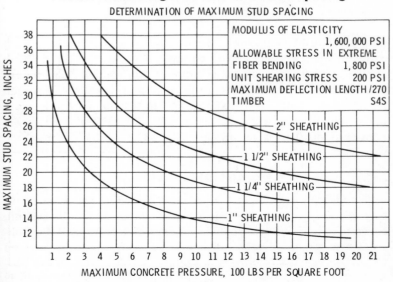

DETERMINATION OF MAXIMUM STUD SPACING

Table 4. Spacing of Wales on Wall Forms

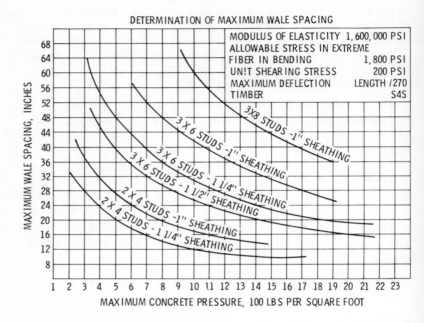

DETERMINATION OF MAXIMUM WALE SPACING

MODULUS OF ELASTICITY 1,600,000 PSI
ALLOWABLE STRESS IN EXTREME
FIBER IN BENDING   1,800 PSI
UNIT SHEARING STRESS   200 PSI
MAXIMUM DEFLECTION   LENGTH /270
TIMBER   S4S

MAXIMUM CONCRETE PRESSURE, 100 LBS PER SQUARE FOOT

tion, with variable spacing of yokes closer together at the bottom and with wider spacing as you go up. Table 5 will assist in determining the yoke spacing for columns. From the left-hand figures of column height, read across to the column for width. Then read up for yoke spacing. For example, a 10-ft. column with a width of 16″ would have yokes spaced at 29″, 30″, 31″, and 31″ from bottom to top. This is illustrated in Fig. 33.

ANCHOR BOLTS

In addition to reinforced bars or mesh placed into concrete forms before pouring, some construction calls for anchor bolts set into concrete. These are protruding and threaded bolts used for subsequent fastening to walls, sills, or other structures to be fastened to the concrete after curing. An example are the bolts shown protruding above the foundation in the illustration shown in Fig. 10.

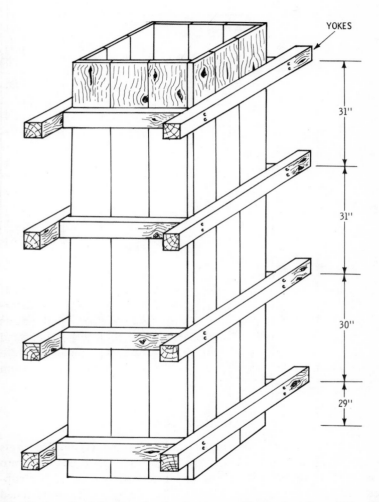

YOKES

31"

31"

30"

29"

Fig. 33. A sectional forming showing the proper yoke spacings for a 18" x 24" column 10 ft. high.

135

Table 5. Yoke Spacing for Columns

LARGEST DIMENSION OF COLUMN IN INCHES

HEIGHT	16"	18"	20"	24"	28"	30"	32"	36"
1'	31"	29"	27"	23"	21"	20"	19"	17"
2'								
3'	31"	28"	26"	23"	21"	20"	19"	17"
4'							18"	17"
5'				23"	20"	19"		15"
6'		28"	26"			18"	17"	12"
7'	30"			22"	18"	18"	13"	11"
8'			24"		15"		12"	10"
9'	29"	26"		16"	13"	12"	10"	8"
10'		20"	19"	14"	12"	10"	10"	8"
11'	21"		16"	13"	10"	9"	8"	7"
12'		18"	15"	12"	9"	8"	8"	6" 6"
13'	20"	16"	14"	11"	9"	8"	7"	6" 6"
14'	18"		14"	10"	8"	7" 7"	6" 6"	
15'	15"	15"	12"	9" 9"	8" 8"	7"	6"	
16'	14"	13"	11"	9" 9"	6" 7"			
17'	13"	12"	11" 11"	8" 8"				
18'	13"	12"	10" 10"	8" 8"				
19'	13"	11"	10" 10"					
20'	12"	11"	9" 10"					

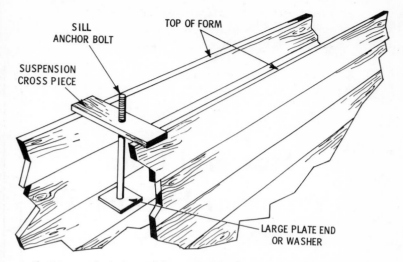

SILL
ANCHOR BOLT

TOP OF FORM

SUSPENSION
CROSS PIECE

LARGE PLATE END
OR WASHER

Fig. 34. A method of suspending a sill anchor bolt for casting in concrete.

Anchor bolts are shown in Fig. 34. There are many types and sizes of anchor bolts, differing only in the manner in which they are held in the concrete. Most types are held in place with a cross piece across the top of the foundation form, and must be precisely placed for proper use later by carpenters for wall sill installation. Their position must not be disturbed during the pouring operation.

137

Hot and Cold
Weather Handling

Placing concrete in hot weather results in detrimental effects that require special precautions for a successful job. Hot weather increases the curing rate to the extent that finishing operations may become very difficult. Because of rapid hydration, the eventual compressive strength of concrete is reduced.

Table 1 shows the relative strength of concrete at high temperatures, as a percent of its strength when cured at 73°F. Note, too, how the early cure rate is higher at higher temperatures. While concrete at 120°F has twice the strength of concrete at 73°F after only one day of curing, the 28-day strength of concrete is less. The one-day curing rate is what is deterimental to screeding and finishing operations.

When concrete is placed in hot weather, more water is needed because of rapid evaporation. Table 2 shows how the water requirement increases with temperature. Remembering the need to maintain a constant water/cement ratio, this condition also means more concrete must be used to maintain good concrete strength when cured. What this statement and chart means is less sand and aggregate is used with higher temperatures to produce a given slump plasticity for a workable concrete.

When concrete is placed in layers, poor adherence of successive layers may result from hardening of the lower layer before the next

Table 1. Relative Final Strength of Concrete at Various Temperatures

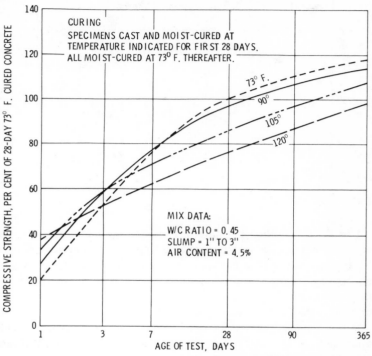

Courtesy Portland Cement Association.

layer is placed. Cracking of the surface is also a result of hot weather placement of concrete. Shrinkage at the surface results from the extra water used, and cracks could occur. This can also result from a cool evening following the placement of concrete during the heat of the day. It is obvious that certain precautions must be taken to prevent the rapid hardening of concrete when placed in hot weather.

PRECAUTIONS

In hot weather, every precaution should be taken to keep the concrete as cool as possible. There are many ways in which this

Table 2. Water Content at Various Temperature Changes

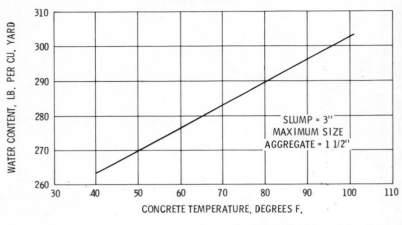

Courtesy Portland Cement Association.

can be done. Cooling the materials making up the concrete is one way of obtaining cooled concrete. Using cold water and cooling the aggregate results in cooled concrete. Table 3 shows the resulting concrete temperature (diagonal lines) with temperatures of the water and the aggregates. This chart was calculated for a given mix, but holds approximately true for all normal mixes.

Water may be cooled by refrigeration or by adding ice. The latter method is usually used in most applications, except where large pouring jobs make the addition of refrigeration equipment economical. When ice is added to water, its weight must be included in calculating the water/cement ratio. The ice must be completely melted by the time it leaves the mixer.

Aggregate should be cooled by pouring water on it while in piled storage, then be kept covered with a tarpaulin which is also kept moist. Water the aggregate again with cold water just before using it. Before the concrete is poured, moisten the earth and forms that hold the concrete. This will reduce the absorption of water from the concrete. Set up windbreaks around the concrete area. This is especially important where high dry winds are present. On very hot days, restrict the pouring of concrete until late in the day when temperatures begin to drop. It may even be advisable to do

Table 3. Effect of Water and Aggregate Temperature

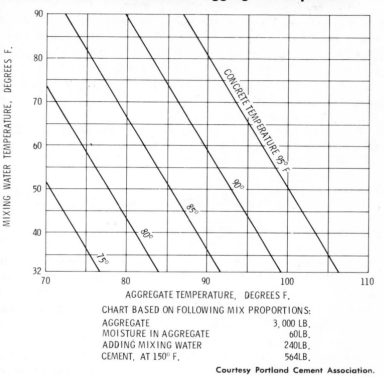

CHART BASED ON FOLLOWING MIX PROPORTIONS:

AGGREGATE	3,000 LB.
MOISTURE IN AGGREGATE	60LB.
ADDING MIXING WATER	240LB.
CEMENT, AT 150° F.	564LB.

Courtesy Portland Cement Association.

the pouring in the evening or at night, especially in arid parts of the country such as the Southwest.

If concrete is ordered from *ready-mix* trucks, be sure they are scheduled for immediate placement, and that you are ready for them the moment they arrive. The truck must not be allowed to stand around in the hot sun while you are doing finishing chores before the placement of the concrete. Be ready to screed (level) the concrete immediately after pouring, and start the finishing as soon as the concrete is able to support the weight of finishing operations.

In hot weather, the use of admixtures to act as hydration retardants is advisable. *Pozzolith* is a favorite admixture for use in hot weather.

141

CURING

Curing should begin immediately after finishing. The first few hours are the most important. If possible, water should be sprinkled onto the concrete before a final curing covering is placed. Keep all wood forms moist for several hours after finishing, and remove them completely within 24 hours if the concrete has hardened enough to be self-supporting. If the forms are left in place, they will tend to absorb water from the concrete.

After 24 hours of moist curing, apply a monomolecular curing film, or cover with plastic sheeting or waterproof paper. The covering should be white to reflect the rays of the drying sun.

COLD WEATHER CONCRETING

Concrete sets slowly in cold weather and delays finishing operations and the removal of forms. If concrete freezes during the early curing period, its strength is adversely affected. Concrete can be placed throughout the winter months if certain precautions are taken. For successful winter work, adequate protection must be provided when temperatures of 40°F or lower occur during placing and during the early curing period.

Plastic concrete must be protected against the disruptive effects of freezing. This danger exists until the degree of saturation has been sufficiently reduced by the withdrawal of mix water in the process of hydration. If no water is available from outside, this reduction will have been accomplished when a compressive strength of about 500 psi has been attained.

Furthermore, protection sometimes must be afforded until the concrete has attained minimum properties required by the service to which it will be exposed. Often, resistance of the surface to damaging effects of saturated freezing and thawing is the governing factor. Sometimes, however, protection to assure freeze-thaw durability may not be adequate for structural safety.

To protect fresh concrete, plans should be made well in advance. Appropriate equipment should be available for heating the concrete materials, for constructing enclosures, and for maintaining favorable temperatures after the concrete is placed.

Table 4. Recommended Concrete Temperature for Cold-Weather Construction

Line	Condition of placement and curing		Thin sections	Moderate sections	Mass sections
1	Min. temp. fresh concrete as	Above 30° F.	60	55	50
2	mixed for weather indicated,	0 to 30° F.	65	60	55
3	deg. F.	Below 0° F.	70	65	60
4	Min. temp. fresh concrete as placed, deg. F.		55	50	45
5	Max. allowable gradual drop in temp. throughout first 24 hours after end of protection, deg. F.		50	40	30

Courtesy Portland Cement Association.

To prevent freezing until protection can be provided, the temperature of concrete, after placing, should not be less than shown in line 4 of Table 4. In addition to the recommended minimum temperatures of concrete after mixing, shown in lines 1, 2, and 3, thermal protection may be required. This is to assure that subsequent concrete temperatures do not fall below the minimums shown in line 4, Table 4 for the periods shown in Table 5 to ensure durability or to develop strength.

Precautions

In placement of concrete in cold weather, four precautions should be observed. Any one or more of these may be followed, depending on the severity of the cold weather.

1. Heat both the water and the ingredients to at least 50°F, but not over 90°F. The net temperature effect is clearly shown in Table 3. Heating the aggregate may require building a shelter over the pile and installing a kerosene-burning space heater in a depression made in the middle of the pile. Since water is more easily heated, aggregate heating is usually avoided unless the air temperature is below freezing.

2. The ground on which the concrete is poured must not be frozen. Otherwise, later thawing will cause some shifting of the earth and possible cracking of the concrete. Placing a sub-base of heated gravel over the ground will thaw it out long enough to pour the concrete.

143

3. Use "High Early Strength" or Type III cement. The faster setting of Type III cement permits an earlier removal of protection against freezing, and earlier finishing.
4. Add 2% (by weight of cement) of calcium chloride. This acts as an accelerator to the curing process. However, chlorides must be avoided if any iron (or steel) or aluminum is embedded in the concrete, since it will attack these metals. Calcium chloride or salt should never be added in great quantities as an antifreeze—this will affect the strength of the concrete. Protection against freezing must be done mechanically.

Table 5. Recommended Duration of Protection for Concrete Placed in Cold Weather

Degree of exposure to freeze-thaw	Normal concrete**	High-early-strength concrete†
No exposure	2 days	1 day
Any exposure	3 days	2 days

Protection for durability at temperature indicated in line 4 of Table 4. Adapted from Recommended Practice for Cold Weather Concreting (ACI 306-66). **Made with Type I, II, or Normal cement. †Made with Type III or High-Early-Strength cement, or an accelerator, or an extra 100 lb. of cement.

Courtesy Portland Cement Association.

HANDLING

Avoid sudden removal of the protection from the warm concrete. The concrete must be allowed to cool slowly. Observe the recommendations of Table 5.

Winter days can be days of low humidity, resulting in rapid evaporation of water from the concrete. Coverings should be used to prevent rapid evaporation of water, as well as quick loss of heat. Usually, the same type of protection will take care of both. Leave forms on an extra day, and cover the slabs with special blankets, or with a thick layer of straw. Never use frozen material as ingredients, or forms containing ice. They must be thawed out by heat. Maintain the temperature of concrete above 50°F for at least three days (during the curing period) for regular type cement, or two days for "High Early Strength" and Type III cement.

CHAPTER 8

Pouring, Finishing, and Curing

The placement of concrete has been treated in previous chapters, but a few reminders at this point are in order.

Forms for footings, walls, etc., should be oiled with commercial form oil or clean motor oil before the concrete is placed so they may be removed easily and used again if necessary. The edge forms, which also serve as leveling screeds, should also be oiled.

Concrete for walls should be placed in 12″ to 18″ layers around the entire wall. The concrete should be spaded with a flat scraper or other thin-bladed tool or mechanically vibrated. This is done to eliminate a condition called "honeycombing" which occurs when coarse aggregate collects at the face of the wall. In inaccessible areas, the forms can be lightly tapped with a hammer to achieve the same result.

Spud vibrators are excellent tools to consolidate fresh concrete in walls and other formed work. The spud vibrator is a metal tube-like device which vibrates at several thousand cycles per minute. When inserted in the concrete for 5 to 15 seconds, the spud vibrator consolidates it and improves the surfaces next to the forms. When pouring flat slabs such as walks, driveways, etc., make the necessary preparations to pour the farthest point first.

The supporting soil for all flat concrete work should be adequately compacted to prevent unequal settling. Prior to placing the concrete, the earth or granular sub-base should be dampened to prevent it from drawing water from the freshly placed concrete. Concrete should never be placed on frozen earth or earth that is flooded with water.

SCREEDING

To screed is to strike-off or level slab concrete after pouring. To do this, and go on with successful finishing, it is well to understand the nature of concrete.

Any craftsman or tradesman should understand the nature and properties of the materials with which he works. A cement mason should understand the nature of concrete, and the different problems in working with concrete made with the various types of portland cement.

Generally, all of the dry materials used in making quality concrete are heavier than water. Thus, shortly after placement, these materials will have a tendency to settle to the bottom and force any excess water to the surface. This reaction is commonly called "bleeding." This bleeding usually occurs with non-air-entrained concrete. It is of utmost importance that the first operations of placing, screeding, and darbying be performed before any bleeding takes place. The concrete should not be allowed to remain in wheelbarrows, buggies, or buckets any longer than is absolutely necessary. It should be dumped and spread as soon as possible and struck off to the proper grade, and then immediately screeded, followed at once by darbying. These last two operations should be performed before any free water has bled to the surface. The concrete should not be spread over a large area before screeding—nor should a large area be screeded and allowed to remain before darbying. If any operation is performed on the surface while the bleed water is present, serious scaling, dusting, or crazing can result. This point cannot be over-emphasized and is the basic rule for successful finishing of concrete surfaces.

The surface is struck off or rodded by moving a straightedge back and forth with a sawlike motion across the top of the forms or screeds. A small amount of concrete should always be kept ahead of the straightedge to fill in all the low spots and maintain a plane surface. Many of the operations described here were also handled in detail in Chapter 3 on Tools, along with illustrations.

TAMPING OR JITTERBUGGING

On some jobs, the next operation is using the hand tamper or jitterbug (Fig. 6, Chapter 3). This tool should be used sparingly and in most cases not at all. If used, it should be used only on concrete having a low slump (1" or less) to compact the concrete into a dense mass. Jitterbugs are sometimes used on industrial floor construction because the concrete for this type of work usually has a very low slump, with the mix being quite stiff and perhaps difficult to work.

The hand tamper or jitterbug is used to force the large particles of coarse aggregate *slightly* below the surface in order to enable the cement mason to pass his darby over the surface without dislodging any large aggregate. After the concrete has been struck off or rodded, and in some cases tamped, it is smoothed with a darby to level any raised spots and fill depressions (Fig. 8, Chapter 3). Long-handled floats of either wood or metal, called *bull floats,* are sometimes used instead of darbies to smooth and level the surface.

FINISHING

When the bleed water and water sheen has left the surface of the concrete, finishing may begin. Finishing may take one or more of several forms, depending on the type of surface desired.

Finishing operations must not be overdone, or water under the surface will be brought to the top. When this happens, a thin layer of cement is also brought up and later, after curing, it becomes a scale which will powder off with usage. Finishing can be done by hand or by rotating power-driven trowels or floats. The size of the job determines the choice, based on economy.

The type of tool used for finishing affects the smoothness of the concrete. A wood float puts a slightly rough surface on the concrete. A steel (or other type metal) trowel or float produces a smooth finish. Extra-rough surfaces are given to the concrete by running a stiff-bristled broom across the top.

Floating

Most sidewalks and driveways are given a slightly roughened surface by finishing with a float. Floats may be small hand-held

Fig. 1. Typical wood float.

Fig. 2. Using the hand wood float from the edge of a slab. Often the worker will work out on the slab on a kneeling board.

tools (Fig. 1), with the work being done while kneeling on a board (Fig. 2), or they may be on long handles for working from the edge. Fig. 3 shows a workman using a long-handled float, and Fig. 4 shows the construction details for making a float.

When working from a kneeling board, the concrete must be stiff enough to support the board and the workman's weight without deforming. This will be within two to five hours from the time the surface water has left the concrete, depending on the type of concrete, any admixtures included, plus weather conditions. Experience and testing the condition of the concrete determines this.

Floating has other advantages. It also embeds large aggregate beneath the surface, removes slight imperfections such as bumps and voids, and consolidates mortar at the surface in preparation for smoother finishes, if desired.

Floating may be done before or after edging and grooving. If the line left by the edger and groover is to be removed, floating

Fig. 3. The long-handled float.

should follow the edging and grooving operation. If the lines are to be left for decorative purposes, edging and grooving will follow floating.

Troweling

Troweling, when used, follows floating. The purpose of troweling is to produce a smooth, hard surface. For the first troweling, whether by power or by hand, the trowel blade must be kept as flat against the surface as possible. If the trowel blade is tilted or pitched at too great an angle, an objectionable "washboard" or

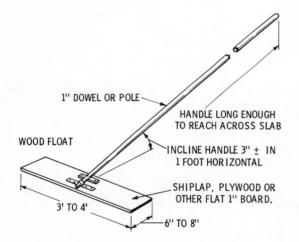

Fig. 4. Construction details for making a long-handled float.

"chatter" surface will result. For first troweling, a new trowel is not recommended. An older trowel which has been "broken in" can be worked quite flat without the edges digging into the concrete. The smoothness of the surface could be improved by timely additional trowelings. There should necessarily be a lapse of time between successive trowelings in order to permit the concrete to increase its set. As the surface stiffens, each successive troweling should be made by a smaller-sized trowel to enable the cement mason to use sufficient pressure for proper finishing.

150

Brooming

For a rough-textured surface, especially on driveways, brooming provides fine scored lines for a better grip for car tires. Brooming lines should always be at right angles to the direction of travel.

Fig. 5. A stiff-bristled broom puts parallel lines in the concrete for a better grip.

For severe scoring, use a wire brush or a stiff-bristled push broom. This operation is done after floating. For a finer texture, such as might be used on a factory floor, use a finer-bristled broom. This operation is best done after troweling to a smooth finish.

Brooming must be done in straight lines (Fig. 5), never in a circular motion. Draw the broom toward you, one stroke at a time, with a slight overlap at the edge of each stroke.

151

Grooving and Edging

In any cold climate there is a certain amount of freezing and thawing of the moist earth under the concrete. When water freezes, it expands. This causes heaving of the ground under the concrete, and this heaving can cause cracking of the concrete in random places.

Sometimes the soil base will settle because all air pockets were not tamped out, or because a leaky water pipe under the soil washed some of it away. A root of a nearby tree under that part of the soil can cause it to lift as the root grows. For all of these reasons

Fig. 6. Cutting a groove in a walk. If any cracking occurs, it will be in the groove where it is less conspicuous.

the concrete can be subjected to stresses which can cause random cracking, even years later.

To avoid random cracking from occurring due to heaving, grooves are cut into the concrete at intervals. These grooves will become the weakest part of the concrete, and any cracking will occur in the grooves. Since, in many cases, heaving or settling cannot be avoided, it is better to have cracks occur in the least conspicuous place possible.

Run a groover across the walk, using a board as a guide to keep the line straight, as shown in Fig. 6. About a 1″ deep groove will be cut, and at the same time, a narrow edge of smoothed concrete will be made by the flat part of the groover.

Fig. 7. An edger rounds off the edge of a walk or driveway.

A rounded edge should be cut along all edges of concrete where it meets the forms, with an edging tool. Running it along the edge of the concrete, between the concrete and the forms, puts a slight round to the edge of the concrete which helps prevent the edges

153

from cracking off, and also gives a smooth-surfaced border. See (Fig. 7).

In the illustration of Fig. 8, masons are putting the finishing touches to a concrete sidewalk. One man is using an edger, while two are floating the surface. In Fig. 9, a floor slab for a home has just been finished. It has a rough texture since the floor will be covered with carpeting when the house is up and ready for occupancy. Water and sewer lines have been laid in position before the concrete was poured.

Fig. 8. Finishing a concrete sidewalk.

FINISHING AIR-ENTRAINED CONCRETE

Air entrainment gives concrete a somewhat altered consistency that requires a little change in finishing operations from those used with non-air-entrained concrete.

Air-entrained concrete contains microscopic air bubbles that tend to hold all of the materials in the concrete (including water)

154

in suspension. This type of concrete requires less mixing water than non-air-entrained concrete, and still has good workability with the same slump. Since there is less water, and it is held in suspension, little or no bleeding occurs. This is the reason for slightly different finishing procedures. With no bleeding, there is no waiting for the evaporation of free water from the surface before starting the floating and troweling operation. This means

Fig. 9. The finished slab of a home under construction. Water and sewer pipes were placed before concrete was poured.

that general floating and troweling should be started sooner— before the surface becomes too dry or tacky. If floating is done by hand, the use of an aluminum or magnesium float is essential. A wood float drags and greatly increases the amount of work necessary to acomplish the same result. If floating is done by power, there is practically no difference between finishing procedures for air-entrained and non-air-entrained concrete, except that floating can start sooner on the air-entrained concrete.

Practically all horizontal surface defects and failures are caused by finishing operations performed while bleed water or excess surface moisture is present. Better results are generally accomplished, therefore, with air-entrained concrete.

CURING

Two important factors affect the eventual strength of concrete:
1. The water/cement ratio must be held constant. This was discussed in detail in previous chapters.
2. Proper curing is important to eventual strength. Improperly cured concrete can have a final strength of only 50% of that of fully cured concrete.

It is hydration between the water and the cement that produces strong concrete. If hydration is stopped due to evaporation of the water, the concrete will become porous and never develop the compressive strength it is capable of producing.

The following relates various curing methods and times compared to the 28-day strength of concrete when moist-cured continuously at 70°F.

1. Completely moist-cured concrete will build to an eventual strength of over 130% of its 28-day strength.
2. Concrete moist-cured for 7 days, and allowed to air dry the remainder of the time, will have about 90% of the strength of example (1) at 28 days and only about 75% of eventual strength.
3. Concrete moist-cured for only 3 days will have about 80% of the 28-day strength of example (1) and remain that way throughout its life.
4. Concrete given no protection against evaporation will have about 52% of 28-day strength, and remain that way.

Curing, therefore, means applying some means of preventing evaporation of the moisture from the concrete. It may take the form of adding water or applying a covering to prevent evaporation, or both.

Curing Time

Hydration in concrete begins to take place immediately after the water and cement are mixed. It is rapid at first, then tapers

156

off as the time goes on. Theoretically, if no water ever evaporates, hydration goes on continuously. Practically, however, all water is lost through evaporation, and after about 28 days, hydration nearly ceases, although some continues for about a year.

Actually, curing time depends on the application, the temperature, and the humidity conditions. Lean mixtures and large massive structures, such as dams, may call for a curing period of a month or more. For slabs laid on the earth, with a temperature around 70°F and humid conditions with little wind, effective curing may be done in as little as 3 days. In most applications, curing is carried for 5 to 7 days.

Table 1. Relative Concrete Strength Versus Curing Method

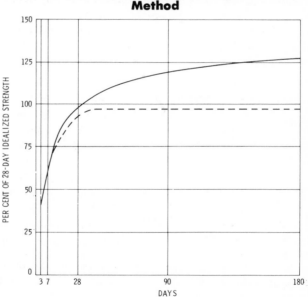

Table 1 shows the relative strength of concrete between an ideal curing time and a practical time. The solid line is the relative strength of concrete kept from any evaporation. Note that its strength continues to increase, but at a rather slow rate with increase in time. The dotted line is the relative strength of concrete that has been cured for 7 days, then allowed to be exposed to free air after that. Strength continues to build until about 28 days, then levels off to a constant value after that. Curing methods should be applied immediately on concrete in forms, and immediately after finishing of flat slabs.

Curing Methods

On flat surfaces such as pavements, sidewalks, and floors, concrete can be cured by *ponding*. Earth or sand dikes around the perimeter of the concrete surface retain a pond of water within the enclosed area. Athough ponding is an efficient method for preventing loss of moisture from the concrete, it is also effective for maintaining a uniform temperature in the concrete. Since ponding generally requires considerable labor and supervision, the method is often impractical except for small jobs. Ponding is undesirable if fresh concrete will be exposed to early freezing.

Continuous *sprinkling* with water is an excellent method of curing. If sprinkling is done at intervals, care must be taken to prevent the concrete from drying between applications of water. A fine spray of water applied continuously through a system of nozzles provides a constant supply of moisture. This prevents the possibility of "crazing" or cracking caused by alternate cycles of wetting and drying. A disadvantage of sprinkling may be its cost. The method requires an adequate supply of water and careful supervision.

Wet coverings such as burlap, cotton mats, or other moisture retaining fabrics are extensively used for curing. Treated burlaps that reflect light and are resistent to rot and fire are available.

Forms left in place provide satisfactory protection against loss of moisture if the top exposed concrete surfaces are kept wet. A soil-soaker hose is an excellent means of keeping concrete wet. Forms should be left on the concrete as long as practicable.

Wood forms left in place should be kept moist by sprinkling, especially during hot, dry weather. Unless wood forms are kept moist, they should be removed as soon as practicable and other methods of curing started without delay.

The application of *plastic sheets* or *waterproof paper* over slab concrete has become one of the most popular methods of curing.

Sprinkle a layer of water over the slab and lay the sheets on top. Tack the edges of the sheets to the edge forms or screeds to keep the water from evaporating. If the sheets are not wide enough to cover the entire area with one piece, use a 12″ overlap between

Fig. 10. Plastic sheeting is a popular covering for curing concrete slabs.

sheets. Use white-pigmented plastic to reflect the rays of the sun, except in cold weather when you want to maintain a warm temperature on the concrete. Waterproof paper is available for the same application. Keep the sheets in place during the entire curing

Fig. 11. A concrete slab with curing compound sprayed on the surface.

period. The photo of Fig. 10 shows plastic sheeting being laid on a newly finished walk.

The use of a liquid *curing compound* that may be sprayed on the concrete is increasing in popularity. It is sprayed on from a hand spray or power spray. It forms a waterproof film on the concrete that prevents evaporation. Its disadvantage is that the film may be broken if the concrete bears the weight of a man or vehicle before the curing period is completed. An advantage is that it may be sprayed on the vertical portions of cast-in-place concrete after the forms are removed.

Fig. 11 shows a slab after applying a curing compound. The black appearance is the color of the curing compound which is sprayed on top of the concrete. A white-pigmented compound is better when the concrete is exposed to the hot sun. Table 2 lists various curing methods, their advantages and disadvantages.

Table 2. Curing Methods

Method	Advantage	Disadvantage
Sprinkling with water or covering with wet burlap.	Excellent results if constantly kept wet.	Likelihood of drying between sprinklings. Difficult on vertical walls.
Straw	Insulator in winter.	Can dry out, blow away, or burn.
Curing compounds.	Easy to apply. Inexpensive.	Sprayer needed; inadequate coverage allows drying out; film can be broken or tracked off before curing is completed; unless pigmented, can allow concrete to get too hot.
Moist earth	Cheap, but messy.	Stains concrete, can dry out, removal problem.

Table 2. Curing Methods (Continued)

Method	Advantage	Disadvantage
Waterproof paper.	Excellent protection, prevents drying.	Heavy cost can be excessive. Must be kept in rolls; storage and handling problem.
Plastic film	Absolutely watertight, excellent protection. Light and easy to handle.	Should be pigmented for heat protection. Requires reasonable care and tears must be patched; must be weighed down to prevent blowing away.

CHAPTER 9

Fancy Finishes

Concrete walks and driveways around a home or modern office building need not be drab in appearance, but may be given simple or involved treatment to take it out of the ordinary. Various patterns may be given to concrete while it is still in its plastic state.

Patterned finishes should be done only by the mason with some experience in this type of work, or someone with the touch of the artist in him. A botched-up pattern design is worse than no pattern at all. Fancy finishes can be applied as soon as the concrete has taken enough set to hold a man on a knee pad, the same time as for standard finishing operations.

SWIRL FLOAT

Perhaps the simplest of finishes with a slight pattern is one that can be made with a long-handled wood float. Best effect is after smooth troweling the surface of the concrete. Allow an hour or so after troweling, then, move the wood float over the concrete surface in overlapping circular motions, or in a overlapping figure eight motion.

The pattern made by this method should not be deep or too obvious. There should be a barely discernible effect of the circles or figure eights showing. The concrete must be allowed to remain undisturbed during the entire curing period. If traffic is allowed on the concrete too early, the pattern will be disturbed.

DESIGNS WITH A BROOM

Patterns like the above, but with deeper indentations, can be made with a stiff-bristled broom. The broom can make other interesting patterns, limited only by one's talent and imagination. The "wavy broom" design is made by drawing the broom across the slab in a left-right-left-right fashion. It will look like a series of waves or modified figure S's.

Each time the broom is drawn across, there must be a slight overlap, and the pattern must be identical with each stroke. As in the case of patterns made with a float, the concrete must be allowed to cure until the marks from the broom are unaffected by traffic on the walk or driveway.

ROCK-SALT TREATMENT

An interesting pockmark effect is obtained by the use of rock salt. Sprinkle rock salt on the surface of the concrete before it is hard, and press the salt into the concrete with a float. After the concrete is thoroughly cured and hard (after 5 days), vigorously wash out the salt with a garden hose. This will leave indents in the concrete where the salt crystals were. This finish is not recommended in climates with frequent freezes in the winter because considerable damage can occur from water freezing and expanding in the holes.

CIRCLES

An interesting pattern design is made with circles of different diameters. Fig. 1 shows a mason impressing circles into concrete that is still plastic. A more effective pattern design is obtained with circles of various diameters interspersed and in random fashion. The devices for making the circles are ordinary household cans from coffee, vegetables, etc. About three sizes of diameters are usually sufficient. The cans used should have been opened with a crank-type or electric can opener and the edges left without burrs.

Press the edge of the cans into the plastic concrete to a depth of about ¼″. Following this, use a rather stiff paint brush to brush

163

Fig. 1. Pressing the open edge of a can into the plastic concrete to make circle patterns.

away concrete edges that may have been lifted with the removal of the can. Use a float or trowel to smooth down any edges that show a feather of concrete lifted from the edges.

FLAGSTONE EFFECT

Two methods may be used to put geometric patterns into the concrete to give it the appearance of laid flagstones. For the first method, obtain some ¼″ × 1½″ wooden lattice stock and cut it into pieces from 4″ to 30″ in length. Use a jigsaw to cut uneven edges in the sides of the wood pieces (Fig. 2), and waterproof the pieces with a couple of coats of varnish. After the concrete has been "screeded" (leveled with a strike board), let it set for a cou-

Fig. 2. Wood lattice strips cut into uneven pieces.

ple of hours until it becomes firm. Lay the wood pieces on the concrete to form geometric patterns which look like flagstones (Fig. 3). Press them into the concrete with a wood float, and at the same time, float the entire surface again (Fig. 4). The top of the wood pieces should be flush with the top surface of the concrete. Now cover the concrete for curing.

Next day, remove the cover and lift the wood strips out with a patching trowel (Fig. 5). After complete curing, the grooves may be left as is or filled in with a contrasting color of mortar; a pure white mortar is very effective. Use white portland cement and white sand to make a white mortar. Before applying the mortar in the grooves, flush out the grooves with water to keep them cool

165

Fig. 3. Laying the wood strips on the concrete in a geometric pattern.

and moist. First make a bonding paint of white portland cement and water proportioned to the consistency of paint. Apply it to the grooves before applying the mortar. This helps bond the mortar to the concrete. Be careful not to let the paint go over the edges of the grooves onto the flat surface of the concrete. Have a bucket of water and sponge handy to wash off any of the paint that bleeds over. Now, apply the mortar into the grooves. You must work fast before the under coat of cement paint dries. It is easier to do this with two men—one applying the coat of cement paint and the other following immediately with the mortar. Trowel the mortar fill flat with the rest of the concrete.

Fig. 4. Pressing the wood strips into the concrete.

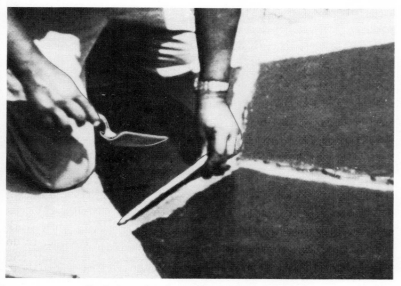

Fig. 5. Removing the wood strips from the concrete.

Fig. 6. Impressing grooves into concrete with a tool.

The other method of obtaining a flagstone pattern is to tool the grooves into the concrete. Make a tool out of a piece of copper tubing (either ½ " or ¾ " in diameter) bent into an S shape. After the concrete has been screeded and after the water sheen has left it, run the curved edge of the copper tubing over the concrete to impress grooves in geometric patterns of random shape and size (Fig. 6). You must not let the concrete get too hard for this method since the tool must push the coarse aggregate aside and the concrete must be quite plastic. Now float-finish the surface in the usual way (Fig. 7).

Run the copper tool over the grooves again after the concrete has set an hour or so, then brush away fine burrs with a paint brush (Fig. 8). Float the surface once more.

LEAF IMPRESSION

This is a special surface which may be used as a border around a patio or along the edges of a garden walk. It is a highly decorative design and adds interest as a conversation piece.

Fig. 7. Float finish concrete after making pattern.

Leaves are taken from local trees, preferably from those on the premises. Immediately after the concrete has been floated and troweled, the leaves should be pressed carefully, stem side down, into the freshly troweled concrete. This is most easily done by using a cement mason's finishing trowel. The leaves should be so completely embedded that they may be troweled over without dislodging them, but no mortar should be deposited over the leaves. After the concrete has set sufficiently, the leaves are removed. Thorough curing is necessary after the concrete has set so the surface will not be marred.

EXPOSED AGGREGATE

A colorful exposed-aggregate surface is often chosen for patios, garden walks, perimeter walls around swimming pools, and driveways—or for any area where a decorative rustic effect is desired. If the surface is ground and polished, it is especially suitable for such places as entrances, interior patios, and recreation rooms.

169

Fig. 8. Brushing out loose pieces of concrete.

The selection of the aggregates is highly important and test panels should be made before the job is started. Colorful gravel aggregate, quite uniform in size—usually ranging from ½" to ¾"—is recommended. Avoid flat, sliver-shaped particles or aggregate less than ½" in diameter because they may not bond properly or may become dislodged during exposing operations. Exposing the aggregate used in ordinary concrete is generally unsatisfactory, since this will just give an unattractive, rough concrete surface.

A 5½- to 6-sack concrete with a maximum slump of 3" should be used. Immediately after the slab has been screeded and darbied, the selected aggregate should be scattered by hand and evenly distributed so that the entire surface is completely covered. The initial embedding of the aggregate is usually done by patting with a darby or the flat side of a 2" × 4" board. After the aggregate is quite thoroughly embedded, and as soon as the concrete will support the weight of a mason on kneeboards, the surface should be hand-floated using a magnesium float or darby. This operation should be performed so thoroughly that aggregate is entirely embedded just beneath the surface. The grout should completely surround and slightly cover all aggregate, leaving no holes or openings in the surface.

Shortly following this floating, a reliable retarder may be sprayed or brushed over the surface, following the manufacturer's recommendations. On small jobs, a retarder may not be necessary. Retarders are generally used on large jobs for better control of exposing operations. Where a retarder has been used, exposing of the aggregate is usually done some hours later by brushing and hosing with water. However, the manufacturer's recommendations should be followed closely.

Whether or not a retarder has been used, the proper time for exposing the aggregate is quite critical. It should be done as soon as the grout covering the aggregate can be removed by simultaneously brushing and hosing with water, yet not overexposing or dislodging the aggregate. If, during exposing, it is necessary for masons to move about on the surface, kneeboards should be used gently. If possible, this should be avoided because of the risk of breaking the aggregate bond.

For interior areas, or where a smooth surface is desired, no retarder is used and exposure of the aggregate is accomplished entirely by grinding. This may be followed by polishing, which will give a surface similar to terrazzo. Because the aggregate completely covers the surface, tooled joints in this type of work are quite impractical. Decorative or control joints are best accomplished by sawing. Control joints should be cut from 4 to 12 hours after the slab is placed. They should be at least one-fifth the depth of the slab. A small-radius edger should be used before and immediately after the aggregate has been embedded to provide a more attractive edge to the slab. Another method of providing control joints is to install permanent strips of redwood before placing the concrete.

In another method of placement, a top course containing the special aggregate and usually 1″ or more thick, is specified. Exposed-aggregate slabs should be cured thoroughly. Care should be taken that the method of curing used does not stain the surface. Straw, earth, and some types of building paper may cause staining.

DRY-SHAKE COLOR SURFACE

This is a colored concrete surface that may be used for showrooms, schools, churches, patios, decorative walks, driveways, or any areas where a colored surface is desired.

It is made by applying a dry-shake material, ready for use, that may be purchased from various reliable manufacturers. Its basic ingredients are mineral-oxide pigment (none other should be used), white portland cement, and specially graded silica sand or fine aggregate. Job selecting, proportioning, and mixing of a dry-shake material is not recommended.

After the concrete has been screeded and darbied, and the free water and excess moisture have evaporated from the surface, the surface should be floated, either by power or by hand float. If done by hand, a magnesium or aluminum float should be used. Before the dry-shake material is applied, preliminary floating should be done to bring up enough moisture for combining with this dry material. Floating also removes any ridges or depressions that might cause variations in color intensity. Immediately following this floating operation, the dry-shake material is shaken evenly by hand over the surface. If too much color is applied in one spot, nonuniformity in color and, possibly, surface peeling will result.

The first application of the colored dry-shake should use about two-thirds of the total amount needed (in lbs. per sq. ft. as specified). In a few minutes this dry material will absorb some moisture from the plastic concrete, and should then be thoroughly floated into the surface, preferably by a power float. Immediately following this, the balance of the dry shake should be distributed evenly over the surface. This should also be thoroughly floated and made part of the surface, making sure that a uniform color is obtained. All tooled edges and joints should be "run" before and after the applications.

Shortly after the final floating operation, the surface should be power troweled. If work is being done by hand, troweling should immediately follow the final floating. After the first troweling, whether by power or by hand, there should be a lapse of time— the length depending on such factors as temperature, humidity, etc., to allow the concrete to increase its set. After this lapse of time, the concrete may be troweled a second time to improve the texture and also produce a denser, harder surface.

Colored slabs, as with other types of freshly placed concrete, must be cured thoroughly. After thorough curing and surface drying, interior surfaces may be given at least two coats of special concrete-floor wax containing the same mineral-oxide pigment used in the dry shake. This wax is also available from various reliable manufacturers. Care should be taken to avoid any staining, such as by dirt or foot traffic, during the curing or drying period and before waxing.

COLORED CONCRETE

Color may be added to a concrete mix and to mortar for laying bricks or concrete blocks. The colors are in the form of mineral pigment which is added to the dry mix and thoroughly mixed in before the water is added. Never add more than 10% of color pigment by weight to the mix. The usual amount is 7 lbs. of color to one 94 lb.-sack of cement for strong colors and down to 1⅓ lbs. of color for pastel shades.

The following are the materials which can be used for colors in concrete or mortar:

Cobalt oxide for blue.
Brown oxide of iron for brown.
Yellow oxide of iron for buffs.
Chromium oxide for green.
Red oxide of iron for red.
Black iron oxide (preferable) or carbon black for gray or slate
    effects (do not use common lamp black).

Use white cement and white sand for a white color. Use white cement and sand with all colors for purer colors. The standard gray cement is preferred for its greater strength, and it should be used if pure color is not important. If you are casting concrete

173

slabs to be laid into a walk or patio later, experiment with both kinds of cement and various colors.

A method that involves more work but is more economical as to color pigment is called the two-course method. Pour regular uncolored concrete into the patio, walk, or block forms, up to about ½" or 1" from the top. Allow it to stiffen a little and the surface water to disappear; then pour the colored concrete over this to the top, and finish in the usual way.

CHAPTER 10

Stucco

The words stucco and plaster are frequently interchanged. The stucco mason is sometimes called a plasterer. More correctly, plaster applies to the covering of inside walls, and stucco to outside walls. Stucco is concrete and has all the qualities of concrete. It is hard, has a long life, is practically impervious to the ravages of weather, and its color is permanent.

INGREDIENTS

Stucco is a cementitious material, containing portland cement, sand, water, and a plasticizer for better workability. The portland cement used in the basic coat is usually Type I, but sometimes Type II. In the Southwest, a finer ground cement is often used and is called plastic cement. The finish coat is made of white portland cement and white sand, with color added.

Water requirements are the same as for regular concrete. It must be free of acids, alkalies, and organic materials. If it is drinkable, the water should be good for stucco. Plasticizers added to make the stucco more workable may be either hydrated lime or short asbestos fibers.

Perhaps one of the most important of the ingredients is the sand. Unlike regular concrete, in which any aggregate under ¼″ is called sand, the sand for stucco must be ⅛″ or less. Even more important is the gradation. Unless the sand is well graded, the voids between the granules will be excessive and require more

Table 1. Sand Gradation for Stucco

Sieve Size No.	Percentage* Passing Each Sieve
4	100
8	100
16	60-90
30	35-70
50	10-30
100	0-5

*The aggregate should have not more than 50% retained between any two consecutive sieves nor more than 25% between the No. 50 and No. 100 sieves.

Courtesy Portland Cement Association.

cement and water. Some complaints have blamed stucco cracking on the fact that too much cement was used, when actually it is the excessive water that is the real culprit. The less water used, the stronger the stucco is. The answer to less water is better gradation of sand to reduce the voids and reduce the amount of cement *and* water needed. Table 1 shows the recommended limits of sand gradation for stucco.

While color may be added to stucco on the job, for better consistency from batch to batch, colored stucco should be purchased from the supplier with the color desired. This is known as "stucco finish." The supplier is able to exercise better control. For the finish coat, all that needs to be added is water.

MIXING

The base coat (under the finish coat) is made of 1 part portland cement and 3 to 5 parts damp loose sand. The range of 3 to 5 parts of sand depends on the gradation, which varies from area to area. Only trial mixes will determine the proper proportion as judged by its workability and ability to adhere without sagging.

If mortar cement or plastic cement is used, no plasticizer will need to be added. For standard portland cement, lime not to exceed 10% by weight of cement or asbestos fibers are added for plasticizers. The amount of asbestos fibers must be determined by trial and error until the right workability is achieved.

For the finish coat, mineral-oxide pigments may be added to the white portland cement. The pigment should be carefully weighed for each batch to maintain consistency of color. As mentioned, it is best to purchase the finish-coat ingredients already prepared, and in the desired color. The dry ingredients should be thoroughly mixed first, then the water added. For uniform quality it is best to use a power mixer. Mixing should be for at least five minutes.

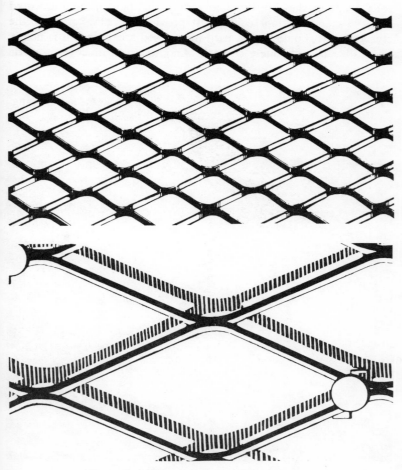

Fig. 1. Expanded metal lath used for stucco lathing.

177

METAL LATHING

For nonporous vertical surfaces, an open mesh metal fabric provides the only satisfactory means of obtaining good adherence of the stucco. The metal reinforcement is available in several forms. Expanded metal lathing is made from flat sheet stock punched to pull out into open diamond-shaped figures. See Fig. 1. Stucco netting (Fig. 2) is wire formed into hexagonal openings and comes in rolls. Some meshes are supplied with V-ribs built in, and are used for soffits and outdoor ceilings. It provides a better grip for the stucco, but it should not be used in exposed areas since it is nearly impossible to completely cover it. Another form of mesh is made

Fig. 2. A wire lathing popular for use on houses. It is known as stucco netting.

of heavy wires interwoven with felt paper and welded where the wires intersect.

The metal lathing described above is used on open frame construction as employed in houses and other small structures. To receive stucco without eventual cracking, the structure must be solid and well braced. The metal lathing must be installed with good rigidity. Otherwise any movement in the structure later could result in cracking the stucco.

Open frame construction is the type using $2'' \times 4''$ vertical studs, usually spaced $16''$ apart. Two systems are used to supply the rigid backing for the stucco, either wood sheathing or longitudinal wires. Frequently, a felt-impregnated wallboard is used in place of the wood sheathing. Tar paper or felt is nailed over the rigid backing and the metal lathing is nailed over the paper or felt.

Fig. 3 shows a sketch of open frame construction using sheathing. The lathing is fastened over the paper or felt with special fur-

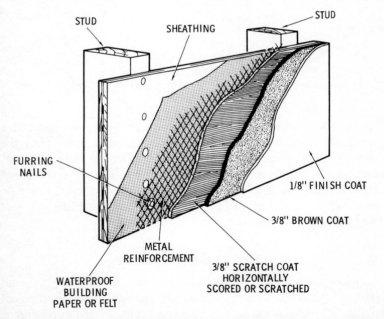

Fig. 3. Open frame construction where wood or impregnated wall board is used as a backing.

179

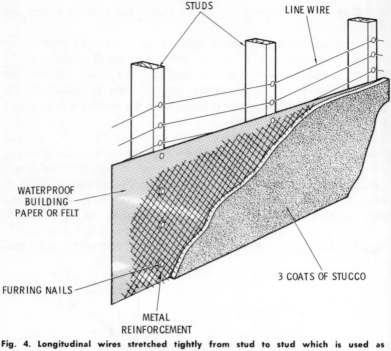

STUDS   LINE WIRE

WATERPROOF
BUILDING
PAPER OR FELT

3 COATS OF STUCCO

FURRING NAILS

METAL
REINFORCEMENT

Fig. 4. Longitudinal wires stretched tightly from stud to stud which is used as a rigid backing.

ring nails which hold the lathing spaced ¼″ from the paper. This permits the stucco to be firmly embedded through and around the mesh for a good mechanical hold. Fig. 4 shows the method of applying horizontal wires instead of sheathing. The wires are nailed to every other stud first, then drawn tight and nailed to the alternate studs. Fig. 5 shows one wall of a home with stucco netting in place. The furring nails are driven into the wall studs.

STUCCO ON MASONRY

Stucco will adhere to masonry walls without the use of metal lathing if the walls are rough enough to provide a good mechanical key, or are porous enough to provide suction. This condition is usually true when applying stucco to concrete-block or cast-in-

Fig. 5. Stucco netting nailed in place on a home.

place concrete walls. Nonporous masonry, such as glazed brick or tile, will require the installation of metal lathing.

Concrete-block walls should be inspected for cleanliness, and any loose dust or grease should be washed off. A water absorption test should be made. If water forms droplets when it is applied, the suction qualities will not be good enough. If water soaks in too quickly, it may absorb water from the stucco too quickly and result in an early stiffening of the stucco. In this case, wet the concrete with water, but do not soak it.

Should there be any question about adherence, bonding agents may be applied. A dash bond of 1 part cement and 1 or 2 parts fine sand can be "thrown" onto the surface with a stiff brush. A special bonding agent, consisting of a water-based emulsion, is available for applying to the surface.

Wall surfaces of old brick, glass block, glazed masonry, metal, or wood do not have the key or porosity to hold onto stucco. Treatment of these is the same as for open frame construction. A tar or felt paper is applied, and metal mesh is fastened over the paper. Special concrete furring nails are used to fasten the metal mesh in place, holding the mesh ¼" from the backing.

181

Old stucco may be refinished by applying a new finishing coat directly over the existing stucco. It must be completely clean to provide good bonding. Stucco should never be applied directly over products containing gypsum, such as plaster. A chemical reaction takes place with the gypsum that destroys the bond.

APPLYING THE BASIC COAT

Most stucco applications are done with three coats—the *scratch* coat, the *brown* coat, and finally the *finish* coat. The scratch coat and the brown coat make up the basic coat. On small jobs, such as homes, hand work is usually employed. For large, multistory buildings, it is more economical to use power equipment which blows the stucco onto the surface. Power blowers are frequently part of the mixer to make the operation continuous.

The first, or scratch coat, should be applied about ½" thick. A mason's metal float is used, plus muscle. It is important to push the stucco well into the mesh so that it surrounds and gets into and behind the metal mesh. The scratch coat is then scored with a scarifier to produce horizontal lines. Fig. 6 shows the sur-

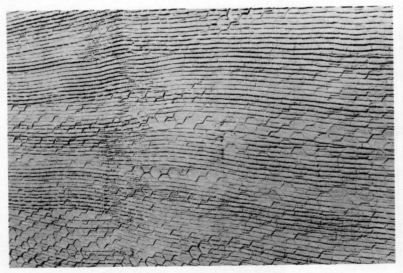

Fig. 6. The first or scratch coat which is applied to the stucco netting.

face texture of a scratch coat. The metal mesh showing through will be covered with the brown coat, although with good practice not more than about 10% of the mesh should be visible.

After some of the moisture has evaporated, but before the scratch sets up too hard, the brown coat is applied. If the scratch coat has been applied against a firm backing, such as sheathing or a masonry wall, the brown coat may be started about 4 or 5 hours later. If the scratch coat has been applied against a wire backing, it should be allowed to take a firm set before applying the brown coat. On the average, this would be about 48 hours later. In this latter case, a fine water spray should be applied to the scratch before putting on the brown coat. The first coat should not be soaked with water, only fine sprayed.

Fig. 7. Applying the brown coat over the scratch or first coat.

183

The second, or brown coat, is applied about ⅜″ thick. It should be placed a full wall at a time, and without interruption. Otherwise workmanship defects in the first coat will show through. The brown coat is applied with a wood float, which leaves a sufficiently rough texture for the finish coat to adhere. Fig. 7 shows the second or brown coat being placed over the scratch coat. Note the straight edge against the wall. It will be used to level the brown coat evenly. The brown coat should be moist-cured for 2 days, then allowed to dry-cure for 5 days before applying the finish coat.

THE FINISH COAT

The finish coat is about ⅛″ thick. It is usually pigmented and often will be given a textured treatment. Only metal-oxide pigments must be used, as described in the previous chapter.

Fig. 8. Applying the finish coat. A straight board is held against the corner for accurate squaring.

Just before applying the finish coat, the brown coat should be evenly dampened to provide uniform suction for better adherence to the finish coat. Fig. 8 shows a mason or plasterer applying the finish coat. His assistant is holding a straight board along the corner to assure a perfectly squared-off corner. The mason is applying a wavy motion to his float to give the finish coat a stippled effect. The effect is shown in Fig. 9.

Fig. 9. The finish coat can be treated in various ways for decorative effect. Shown here is a stipled surface.

A variety of decorative effects may be attained on the finish coat. The stippled effect shown in Fig. 9 may be altered by running a float over the high spots of the raised stucco. A thin coat of finish stucco may be "thrown" onto the coat with a brush to produce a pebble-like effect. The coat may be scored with a broom or other scratching tool. A sandy texture is obtained by rubbing the finish with a float surfaced with soft material, such as rubber, plastic foam, or carpet.

Exposed aggregate, similar to that for concrete walks, may be deposited on the finish stucco while it is still plastic. This is called "rock dash" or "marblecrete." Colored marble chips or small clean

185

rocks are literally thrown at the stucco by hand, or placed into position. Electrically or pneumatically driven guns are available for throwing small rocks against the stucco. They do a more even job and do it faster.

CURING

Since stucco is actually concrete, even though applied to a vertical wall, it must be properly cured the same as concrete, as described in the previous chapters. It is important for hydration to take place without rapid loss of water through evaporation.

Curing of stucco is best done with a garden hose. Moisten the stucco with a fine spray of the hose regularly for about 5 days. The stream must not have high pressure or some of the plastic stucco may be displaced. Where hot dry winds are encountered, hang tarpaulin or plastic sheets over the surface to reduce the drying effects of the winds.

CHAPTER 11

Concrete Block

Fast replacing the common brick, concrete masonry blocks are adding an exciting new look to the field of available construction material. Made of cast concrete, many of them are even less expensive than face brick and other forms of construction. They have unusual strength and are kept from being too heavy by casting them with hollow inner cores.

Used originally for economy, the concrete block is now available in a variety of forms and colors, thus making them a distinctive material appropriate for prestige construction, including homes shown in (Fig. 1).

BLOCK SIZES

Concrete blocks are available in a variety of sizes and shapes. Fig. 2 shows some of the sizes in common use. They are all sized on the basis of multiples of 4″. The fractional dimensions shown allow for the mortar (Fig. 3). Some concrete blocks are poured concrete made of standard cement, sand, and aggregate. An 8″ × 8″ × 16″ block weighs about 40 to 50 lbs. Some use lighter natural aggregates, such as volcanic cinders or pumice, and some are manufactured aggregates such as slag, clay, or shale. These blocks weigh about 25 to 35 lbs.

In addition to the hollow-core types shown, concrete blocks are also available in solid forms. In some areas they are available in other sizes than those shown. Many of the same type have half the

187

(A) Using grille block in a partition wall.

(B) Using standard block and raked joints in a partition wall.

Fig. 1. Examples of concrete block used in home construction.

height, normally 4″, although actually 3⅝″ to allow for mortar. The 8″ × 8″ × 16″ stretcher (center top illustration of Fig. 2) is most frequently used. It is the main block in building a yard wall or a building wall. Corner or bullnose blocks with flat finished ends are used at the corners of walls. Others have special detents for window sills, lintels, door jambs, etc.

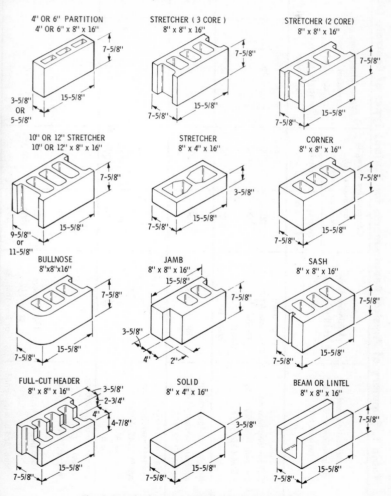

Fig. 2. Standard size and shapes of concrete blocks.

189

Table 1. Summary of Physical Requirements for Various Types of Concrete Masonry Units

Specification, serial designation, and latest revised date	Minimum face-shell thickness, in.	Compressive strength, minimum, psi, average gross area		Water absorption, maximum, lb. per cu. ft. of concrete, average of 5 units	Moisture content, maximum, percent of total absorption, average of 5 units
		Average of 5 units	Individual unit		
Hollow load-bearing concrete masonry units ASTM C90, 1952	1¼ or over: Grade A[a,c] Grade B[b,c] Under 1¼ and over ¾	1000 700 1000	800 600 800	15 — 15	40 40 40
Hollow non-load-bearing concrete masonry units ASTM C129, 1952	Not less than ½	350	300	—	40
Solid load-bearing concrete masonry units ASTM C145, 1952[d] Grade A Grade B	— —	1800 1200	1600 1000	15 15	40 40
Concrete units; masonry, hollow Federal SS-C-621, 1935				Average of 3 units	Average of 3 units
Load-bearing units	1¼ or more ¾ to 1¼	700 1000	600 800	16 16	40 40
Non-load-bearing units	Not less than ¾	350	—	—	40

	Compressive strength, minimum, psi, average gross area (brick flatwise)		Modulus of rupture, minimum, psi, (brick flatwise)		Water absorption, maximum		Moisture content, maximum, percent of total absorption
	Average of 5 brick	Individual	Average of 5 brick	Individual	Average of 5 brick	Individual	
Concrete building brick ASTM C55, 1952					15 lb. per cu. ft.		
Grade A[e]	2500	2000	—	—		—	40
Grade B[f]	1500	1250	—	—		—	40
Brick; concrete Federal SS-B-663, 1932							
H—Hard	—	—	600+	400	7 oz.	9.5 oz.	30
M—Medium	—	—	450-600	300	8 oz.	10 oz.	30
S—Soft	—	—	300-450	200	no limit	no limit	30

[a] For use in exterior walls below grade and for unprotected exterior walls above grade that may be exposed to frost action.

[b] For general use above grade in walls not subjected to frost action or where protected from the weather with two coats of portland cement paint or other satisfactory waterproofing treatment approved by the purchaser.

[c] Regardless of the grade of unit used, protective coatings such as portland cement paint may be desirable on exterior walls for waterproofing purposes. In this connection purchasers should be guided by local experience and the manufacturer's recommendations.

[d] Units with 75 percent or more net area. The classification is based on strength.

[e] Brick intended for use where exposed to temperatures below freezing in the presence of moisture.

[f] Brick intended for use as back-up or interior masonry.

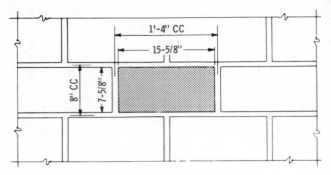

Fig. 3. Illustrating block size to allow for mortar joints.

Fig. 4. Split block laid up in a lattice-like pattern adds designing elegance.

Compressive strength is a function of the face thickness. Concrete blocks vary in thickness of the face, depending on whether they are to be used for non-load-bearing walls, such as yard walls, or load-bearing walls, such as for buildings. Table I lists the grade and *ASTM* specifications of concrete block for various applications.

DECORATIVE BLOCK

In addition to standard rectangular forms, concrete masonry blocks are now being made in usual designs and with special cast-in colors and finishes to make them suitable architectural designs for both indoor and outdoor construction. A few decorative blocks are listed below.

Fig. 5. A wall of slump block adds rustic charm and is especially suitable for ranch-type homes.

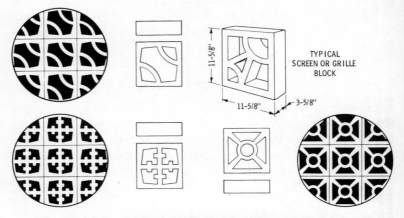

Fig. 6. Examples of grille blocks. The actual patterns available varies with processors in different areas.

Split Block

Resembling natural stone, split block is made from standard 8″ thick pieces split by the processor into 4″ thick facing blocks. The rough side faces out. Split block is usually gray in color, but some have red, yellow, buff, or brown colors made as an integral part of the cast concrete.

Split block is especially handsome as a low fence. By laying it up in lattice-like fashion, it makes a handsome carport, keeping out the weather but allowing air and light to pass through. (Fig. 4).

Slump Block

When the processor uses a mix that slumps slightly when the block is removed from the mold, it takes on an irregular appearance like old-fashioned hand molding. Slump block strongly resembles adobe or weathered stone; and it, too, is available with integral colors and is excellent for ranch-style homes, fireplaces, and garden walls (Fig. 5).

Grille Blocks

Some of the most attractive of the new concrete blocks are grille blocks which come in a wide variety of patterns, a few of which are

Fig. 7. Outside and inside view of a grille block wall.

Fig. 8. Large expansion of glass in home construction can be given some privacy without cutting off all light.

shown in Fig. 6. In addition to its beauty, the grille block provides the practical protection of a concrete wall, yet allows some sun and light to enter (Fig. 7). They are especially useful in cutting the effects of heavy winds without blocking all circulation of air for ventilation. They are usually 4″ to 6″ thick and have faces that are 12″ or 16″ square.

Screen Block

Similar to grille block but lighter in weight and more open, screen block is being used more and more as a facing for large window areas. In this way, beauty and some temperature control is added. They protect the large window panes from icy winter blasts and strong summer sun, yet provide privacy and beauty to the

Fig. 9. An attractive garden wall with a symetrical pattern.

home. Special designs are available, or they may be made up by laying single-core standard block on its side as shown in Fig. 8.

Patterned Block

Solid block may also be obtained with artistic patterns molded in for unusual effects both indoors and out. Some carry the trade names of *Shadowal* and *Hi-Lite*. The first has depressed diagonal recessed sections, and the second has raised half-pyramids. Either can be placed to form patterns of outstanding beauty.

Special Finishes

Concrete block is now being produced by some manufacturers with a special bonded-on facing to give it special finishes. Some are made with a thermosetting resinous binder and glass silica sand which give a smooth-faced block. A marbleized finish is being produced by another manufacturer; a vitreous glaze is still another.

197

Some blocks may be obtained with a striated bark-like texture. Blocks with special aggregate can be found—some ground down smooth for a terrazo effect.

STANDARD CONCRETE BLOCK

Standard concrete blocks with hollow cores can make handsome walls, depending on how they are laid and on the sizes chosen. Fig. 9 shows a solid high wall of standard block which provides maximum privacy. A few are laid hollow core out for air circulation. Fig. 10 is standard block with most of the blocks laid sideways to expose the cores. Fig. 11 shows single-cored, thin-edged concrete blocks used to support a sloping ground level. It prevents earth runoff and adds an unusual touch of beauty.

Fig. 10. Standard double-core block laid with cores exposed for ventilation.

Fig. 11. Single-core corner blocks used in a garden slope.

Fig. 12. Solid concrete blocks used as a patio floor.

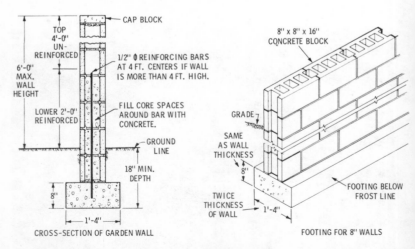

Fig. 13. Extra large precast concrete slabs used to make an attractive walk.

CAP BLOCK

TOP
4'-0"
UN-
REINFORCED

6'-0"
MAX.
WALL
HEIGHT

LOWER 2'-0"
REINFORCED

1/2" Ø REINFORCING BARS
AT 4 FT. CENTERS IF WALL
IS MORE THAN 4 FT. HIGH.

FILL CORE SPACES
AROUND BAR WITH
CONCRETE.

GROUND
LINE

18" MIN.
DEPTH

8"

1'-4"

CROSS-SECTION OF GARDEN WALL

8" x 8" x 16"
CONCRETE BLOCK

GRADE

SAME
AS WALL
THICKNESS

8"

TWICE
THICKNESS
OF WALL

1'-4"

FOOTING BELOW
FROST LINE

FOOTING FOR 8" WALLS

Fig. 14. Cross-section end view of a simple block wall. Vertical reinforcement rods
are placed in the hollow cores at various intervals.

In Fig. 12, precast concrete block of standard size is laid like brickwork for use as a patio floor. Level the earth and provide a gentle slope to allow for rain runoff. Put in about a 2″ layer of sand, and lay the blocks in a two-block criss-cross pattern. Leave a thin space between blocks and sweep sand into the spaces after the blocks are down. Fig. 13 shows a walk made of precast concrete blocks of extra large size. These are not standard, but they are available.

WALL THICKNESS

Garden walls under 4 ft. high can be as thin as 4″, but it is best to make them 8″ thick. Walls over 4 ft. high must be at least 8″ thick to provide sufficient strength.

Fig. 15. Illustrating vertical reinforcement rods through double-thick column blocks.

A wall up to 4 ft. high needs no reinforcement. Merely build up the fence from the foundation with block or brick, and mortar. Over 4 ft., however, reinforcement will be required, and the fence should be of block, not brick. As shown in the sketch of Fig. 14, set ½ " diameter steel rods in the poured concrete foundation at 4-ft. centers. When you have laid the blocks (with mortar) up to the level of the top of the rods, pour concrete into the hollow cores around the rods. Then continue on up with the rest of the layers of blocks.

In areas subject to possible earthquake shocks or extra high winds, horizontal reinforcement bars should also be used in high walls. Use No. 2 (¼ ") bars or special straps made for the purpose. Fig. 15 is a photograph of a block wall based on the sketch of Fig. 14. The foundation is concrete poured in a trench dug out of the ground. Horizontal reinforcement is in the concrete, with vertical members bent up at intervals. High column blocks are laid (16" × 16" × 8") at the vertical rods. The columns are evident in the finished wall of Fig. 16.

Fig. 16. A newly finished concrete block wall. Note reinforcement columns at various intervals.

Load bearing walls are those used as exterior and interior walls in residential and industrial buildings. Not only must the wall support the roof structure, but it must bear its own weight. The greater

Table 2. Allowable Height & Minimum Nominal Thickness for Concrete Masonry Bearing Walls

For buildings up to three stories in height with walls of hollow or solid concrete masonry units. For buildings over three stories, see ASA Code. (Thickness in inches)

Wall	Residential			Nonresidential			Cavity residential			Cavity nonresidential		
	one-story	two-story	three-story	one-story	two-story	three-story	one-story	two-story	three-story	one-story	two-story	three-story
3rd story	—	—	8	—	—	12 / 8^d	—	—	10	—	—	12 / 10^f
2nd story	—	8	8	—	12 / 8^d	12	—	10	10	—	12 / 10^f	12
1st story	8 / 6^a	8	8	12 / 8^d	12	12 / 16^c	10	10	12 or 10^e	12 / 10^f	12	12^g
Basement or foundation	12 or 8^b			12 or 8^b	12	12 or 16	12, 10 or 8^h (not cavity)			12 or 10 (not cavity)		

a May be 6″ for one one-story single-family dwellings and one-story private garages when not more than 9 ft. in height, with an allowance of 6 ft. additional for any gable.

b When foundation wall does not extend more than 4 ft. into the ground the wall may be 8″ thick. With special approval of the building official this depth may be extended to 7 ft. where soil conditions warrant such an extension. In no case shall the total height of 8″ or 10″ concrete masonry walls, including the foundation wall, exceed 35 ft.

c Must be at least 16″ if the total height of the first, second and third stories above the foundation wall, or from a girder or other intermediate supports, is more than 35 ft.

d Top story may be 8″ when not more than 12 ft. in height and roof beams are horizontal and total height of masonry wall is not more than 35 ft.

e In no case shall total height of a 10″ cavity wall above the foundation wall exceed 25 ft.

f Top story may be 10″ when not more than 12 ft. in height and roof beams are horizontal and total weight of wall is not more than 35 ft.

g In no case shall total height of a cavity wall exceed 35 ft. above the foundation wall, regardless of thickness.

h May be 8″ for 1½-story single-family dwellings having a maximum height, including the gable, of not over 20 ft. and having nominal 10″ cavity walls. Such 8″ foundation walls shall be corbelled to provide a bearing the full thickness of the wall above. Total projection not to exceed 2″ with top course a full header course not higher than the bottom of the floor joists. Individual projections in corbelling shall not be more than one-third the height of the unit.

203

the number of stories in the building, the greater the thickness the lower stories must be to support the weight of the concrete blocks above it, as well as roof structure. Table 2 shows wall thicknesses recommended for buildings of various heights as established by the *American Standard Building Code*.

FOUNDATIONS

Any concrete-block or brick wall requires a good foundation to support its weight and prevent any position shift that may produce cracks. Foundations or footings are concrete poured into forms or trenches in the earth. Chapter 6 describes forms and their construction for footings. For non-load-bearing walls, such as yard fences for example, an open trench with smooth sides is often satisfactory. Recommended footing depth is 18″ below the grade level. In areas of hard freezes, the footing should start below the frost line.

Fig. 17. A garden wall of concrete blocks. The columns are spaced 15 ft. apart.

Footings or foundations must be steel reinforced, with reinforcing rods just above the bed level. Reinforcing rods should be bent to come up vertically at regular intervals into the open cores of column sections of walls, where used, or regular sections of the wall if columns are not used. Columns of double thickness are recommended where the wall height exceeds 6 ft. Even lower height walls are better if double-thick columns are included. (See Fig. 17). The earth bed below the foundation must be well-tamped and include a layer of sand for drainage.

MORTAR

Mortar bonds the masonry units together to form a strong durable wall. The mortar must be chemically stable and resist rain penetration and damage by freezing and thawing. Mortar must have sufficient strength to carry all loads applied to the wall for the life of the building with a minimum of maintenance.

Mortar is widely used in home construction for all types of masonry walls. Masonry cement eliminates the need to stockpile and handle extra material and reduces the chance of improper on-the-job proportioning. Consistent mortar color in successive batches is easy to obtain when using masonry cement.

Table 3. Recommended Mortar Mix

Type of service	Cement	Mortar sand in damp, loose condition
For ordinary service	1 masonry cement*	2¼ to 3
Subject to extremely heavy loads, violent winds, earthquakes, or severe frost action. Isolated piers.	1 masonry cement* plus 1 portland cement	4½ to 6

*ASTM Specification C91, Type II.

Table 3 gives the recommended mortar mixes. Water is added until the mortar is plastic and handles well under a trowel. Mortar should be machine-mixed whenever practical. Masonry cement contains an air-entraining agent that causes the formation of tiny air bubbles in the mortar. These bubbles make the mortar more workable when plastic, slow the absorption of water by the masonry unit, and reduce the possibility of weather damage.

BUILDING WITH CONCRETE BLOCKS

Proper construction of concrete-block walls, whether for yard fencing or building structures, requires proper planning. Standard concrete blocks are made in 4″ modular sizes. Their size allows for a ⅜″ thick mortar joint. By keeping this in mind, the width of a

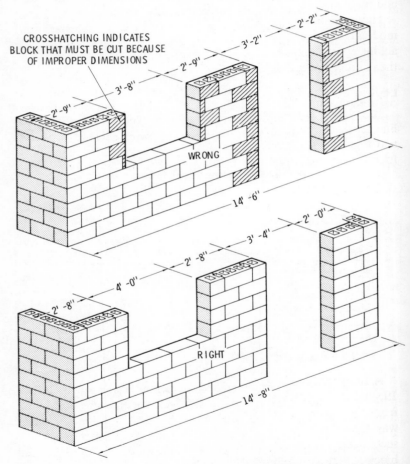

Courtesy Portland Cement Association.

Fig. 18. Right and wrong way to plan door and window openings in block walls.

206

wall and openings for windows and doors may be planned without the need for cutting any of the blocks to fit.

Fig. 18 shows the right and wrong way to plan for openings. The illustration on the left did not take into account the 4″ modular concept, and a number of blocks must be cut to fit the window and door opening. The illustration on the right shows correct planning, and no blocks need to be cut for the openings.

Having established the length of a wall on the basis of the 4″ modular concept (total length should be some multiple of 4″), actual construction begins with the corners. Stretcher blocks are then laid between the corners.

Laying Block at Corners

In laying up corners with concrete masonry blocks, place a taut line all the way around the foundation with the ends of the string tied together. It is customary to lay up the corner blocks, three or four courses high, and use them as guides in laying the walls.

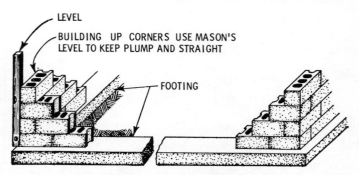

Fig. 19. Laying up corners when building with concrete masonry block units.

A full width of mortar is placed on the footing, as shown in Fig. 19, and the first course is built two or three blocks long each way from the corner. The second course is half a block shorter each way than the first course; the third, half a block shorter than the second, etc. Thus, the corners are stepped off until only the corner block is laid. Use a line, and level frequently to see that the blocks are laid straight and that the corners are plumb. It is customary that such special units as corner blocks, door and window jamb

207

blocks, fillers, veneer blocks, etc., be provided prior to commencing the laying of the blocks.

Building the Wall Between Corners

In laying walls between corners, a line is stretched tightly from corner to corner to serve as a guide (Fig. 20). The line is fastened to nails or wedges driven into the mortar joints so that, when stretched, it just touches the upper outer edges of the block laid in

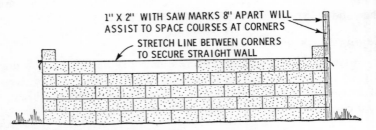

Fig. 20. Showing procedure in laying concrete block walls.

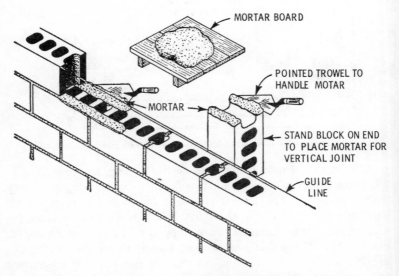

Fig. 21. The usual practice in applying mortar to concrete blocks.

the corners. The blocks in the wall between corners are laid so that they will just touch the cord in the same manner. In this way, straight horizontal joints are secured. Prior to laying up the outside wall, the door and window frames should be on hand to set in place as guides for obtaining the correct opening.

Applying Mortar to Blocks

The usual practice is to place the mortar in two separate strips, both for the horizontal or bed joints and for the vertical or end joints, as shown in Fig. 21. The mortar is applied only on the face shells of the block. This is known as *face-shell bedding*. The air spaces thus formed between the inner and outer strips of mortar help produce a dry wall.

Masons often stand the block on end and apply mortar for the end joint as shown in Fig. 21. Sufficient mortar is put on to make sure that all joints will be well filled. Some masons apply mortar on the end of the block previously laid as well as on the end of the

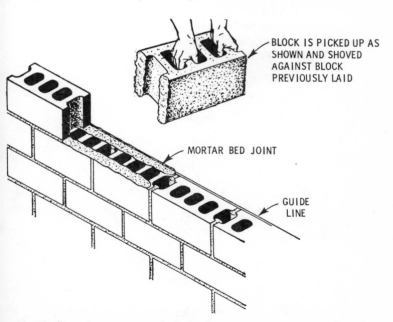

BLOCK IS PICKED UP AS SHOWN AND SHOVED AGAINST BLOCK PREVIOUSLY LAID

MORTAR BED JOINT

GUIDE LINE

Fig. 22. Illustrating common method of picking up and setting concrete blocks.

block to be laid next to it to make certain that the vertical joint will be completely filled.

Placing and Setting Blocks

In placing, the block which has mortar applied to one end is picked up, as shown in Fig. 22, and shoved firmly against the block previously placed. Note that mortar is already in place in the bed or horizontal joints.

Mortar squeezed out of the joints is carefully scraped off with the trowel and applied on the other end of the block or thrown back onto the mortar board for later use. The blocks are laid to touch the line and are tapped with the trowel to get them straight and level as shown in Fig. 23. In a well-constructed wall, mortar joints will average ⅜″ thick. The illustration in Fig. 24 shows a mason building up a concrete-block wall.

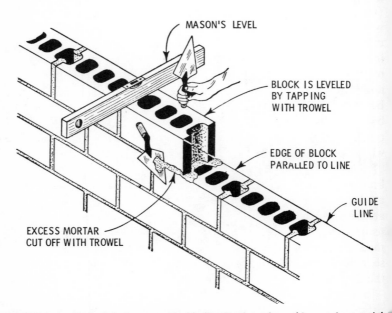

Fig. 23. A method of laying concrete blocks. Good workmanship requires straight courses with the face of the wall plumb and true.

Building Around Door and Window Frames

There are several acceptable methods of building door and window frames in concrete masonry walls. One method used is to set the frames in the proper position in the wall. The frames are then plumbed and carefully braced, after which the walls are built up against them on both sides. Concrete sills may be poured later.

The frames are often fastened to the walls with anchor bolts passing through the frames and embedded in the mortar joints. Another method of building frames in concrete masonry walls is to build openings for them, using special jamb blocks as shown in Fig. 25. The frames are inserted after the wall is built. The only advantage of this method is that the frames can be taken out without damaging the wall, should it ever become necessary.

Placing Sills and Lintels

Building codes require that concrete-block walls above openings shall be supported by arches or lintels of metal or masonry (plain or reinforced). Arches and lintels must extend into the walls not less than 4″ on each side. Stone or other nonreinforced masonry lintels should not be used unless supplemented on the inside of the wall with iron or steel lintels. Fig. 26 illustrates typical methods of inserting concrete reinforced lintels to provide for door and window openings. These are usually prefabricated, but may be made up on the job if desired. Lintels are reinforced with steel bars placed 1½″ from the lower side. The number and size of reinforcing rods depend upon the width of the opening and the weight of the load to be carried.

Sills serve the purpose of providing watertight bases at the bottom of wall openings. Since they are made in one piece, there are no joints for possible leakage of water into walls below. They are sloped on the top face to drain water away quickly. They are usually made to project 1½″ to 2″ beyond the wall face, and are made with a groove along the lower outer edge to provide a drain so that water dripping off the sill will fall free and not flow over the face of the wall causing possible staining.

Slip sills are popular because they can be inserted after the wall proper has been built, and therefore require no protection during construction. Since there is an exposed joint at each end of the sill,

(A) Several blocks are receiving mortar on the end.

(B) Blocks are tapped into position.

Fig. 24. Actual construction

special care should be taken to see that it is completely filled with mortar and the joints packed tight.

Lug sills project into the masonry wall (usually 4″ at each end.) The projecting parts are called *lugs*. There are no vertical mortar joints at the juncture of the sills and the jambs. Like the slip sill, lug sills are usually made to project from 1½″ to 2″ over the face of the wall. The sill is provided with a groove under the lower outer edge to form a drain. Frequently, they are made with washes at either end to divert water away from the juncture of the sills and the jambs. This is in addition to the outward slope on the sills.

At the time lug sills are set, only the portion projecting into the wall is bedded in mortar. The portion immediately below the wall opening is left free of contact with the wall below. This is done in case there is minor settlement or adjustments in the masonry work during construction, thus avoiding possible damage to the sill during the construction period.

(C) Excess mortar is removed and alignment checked.

of a concrete block wall.

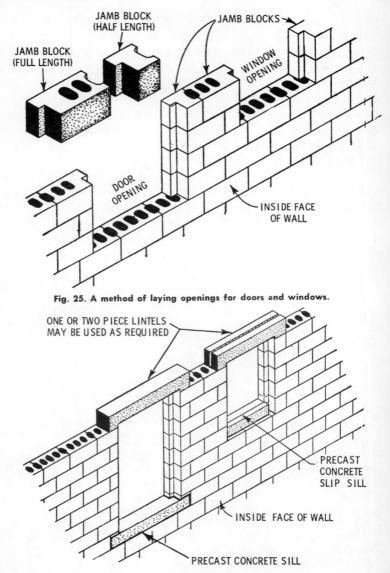

Fig. 25. A method of laying openings for doors and windows.

Fig. 26. A method of inserting precast concrete lintels and sills in concrete block wall construction.

214

BASEMENT WALLS

Basement walls shall not be less in thickness than the walls immediately above them, and not less than 12″ for unit masonry walls. Solid cast-in-place concrete walls are reinforced with at least one ⅜″ deformed bar (spaced every 2 ft.) continuous from the footing to the top of the foundation wall. Basement walls with 8″ hollow concrete blocks frequently prove very troublesome. All hollow block foundation walls should be capped with a 4″ solid concrete block, or else the core should be filled with concrete.

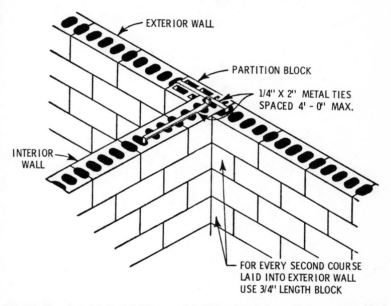

Fig. 27. Showing detail of joining an interior and exterior wall in concrete block construction.

BUILDING INTERIOR WALLS

Interior walls are built in the same manner as exterior walls. Load-bearing interior walls are usually made 8″ thick; partition walls that are not load bearing are usually 4″ thick. The recommended method of joining interior load-bearing walls to exterior walls is illustrated in Fig. 27.

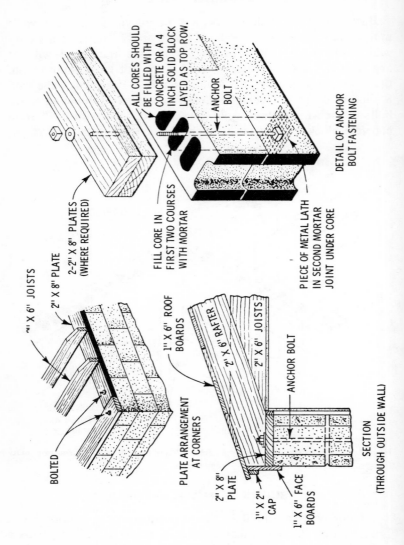

ALL CORES SHOULD BE FILLED WITH CONCRETE OR A 4 INCH SOLID BLOCK LAYED AS TOP ROW.

ANCHOR BOLT

DETAIL OF ANCHOR BOLT FASTENING

FILL CORE IN FIRST TWO COURSES WITH MORTAR

PIECE OF METAL LATH IN SECOND MORTAR JOINT UNDER CORE

2-2" X 8" PLATES (WHERE REQUIRED)

2" X 8" PLATE

'" X 6" JOISTS

1" X 6" ROOF BOARDS

2" X 6" RAFTER

2" X 6" JOISTS

ANCHOR BOLT

BOLTED

PLATE ARRANGEMENT AT CORNERS

2" X 8" PLATE

1" X 2" CAP

1" X 6" FACE BOARDS

SECTION (THROUGH OUTSIDE WALL)

Fig. 28. Showing details of the methods used to anchor sills and plates to concrete block walls.

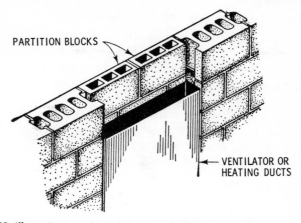

Fig. 29. Illustrating a method of installing ventilating and heating ducts in concrete block walls.

BUILDING TECHNIQUES

Sills and plates are usually attached to concrete block walls by means of anchor bolts, as shown in Fig. 28. These bolts are placed in the cores of the blocks, and the cores filled with concrete. The bolts are spaced about 4 ft. apart under average conditions. Usually ½″ bolts are used and should be long enough to go through two courses of blocks and project through the plate about 1″ to permit use of a large washer and anchor bolt nut.

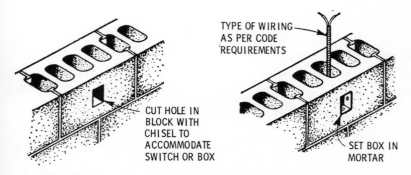

Fig. 30. Method showing the installation of electrical switches and outlet boxes in concrete block walls.

217

Installation of Heating and Ventilating Ducts

These are provided for as shown on the architect's plans. The placement of the heating ducts depends on the type of wall— whether it is load bearing or not. A typical example of placing the heating or ventilating ducts in an interior concrete masonry wall is shown in Fig. 29.

Interior concrete-block walls which are not load bearing, and which are to be plastered on both sides, are frequently cut through to provide for the heating duct, the wall being flush with the ducts on either side. Metal lath is used over the ducts.

Electrical Outlets

These are provided for by inserting outlet boxes in the walls, as shown in Fig. 30. All wiring should be installed to conform with the requirements of the *National Electrical Code* and local codes in the area.

Fill Insulation

In masonry construction, insulation is provided by filling the cores of concrete block units in all outside walls with granulated insulation as construction proceeds.

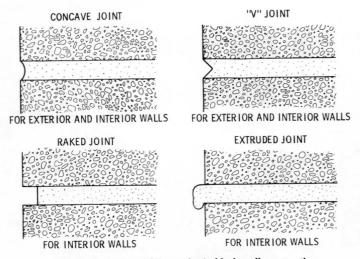

CONCAVE JOINT

"V" JOINT

FOR EXTERIOR AND INTERIOR WALLS

FOR EXTERIOR AND INTERIOR WALLS

RAKED JOINT

EXTRUDED JOINT

FOR INTERIOR WALLS

FOR INTERIOR WALLS

Fig. 31. Four joint styles popular in block wall construction.

Flashing

Adequate flashing with rust- and corrosion-resisting material is of the utmost importance in masonry construction, since it prevents water from getting through the wall at vulnerable points. Points requiring protection by flashing are:

1. Tops and sides of projecting trim under coping and sills.
2. At the intersection of a wall and the roof.
3. Under built-in gutters.
4. At the intersection of a chimney and the roof.
5. At all other points where moisture is likely to gain entrance.

Flashing material usually consists of No. 26 gauge (14-oz.) copper sheet or other approved noncorrodible material.

TYPES OF JOINTS

The concave and V joints are the best for most areas. Fig. 31 shows four popular joints. While the raked and the extruded styles

Fig. 32. Brick wall with extruding joint construction.

Fig. 33. Raking a concave joint with a round S-shaped tool.

are recommended for interior walls only, they may be used outdoors in warm climates where rains and freezing weather are at a minimum. In climates where freezes can take place, it is important that no joint permits water to collect.

In areas where the raked joint can be used, you may find it looks handsome with slump block. The sun casts dramatic shadows on this type of construction. Standard blocks with extruded joints have a rustic look, and make a good background for ivy and other climbing plants (Fig. 32).

Tooling the Joints

Tooling consists of compressing the squeezed-out mortar of the joints back tight into the joints and taking off the excess mortar. The tool should be wider than the joint itself (wider than ½″). You can make an excellent tooling device from ¾″ copper tubing bent into an S-shape. By pressing the tool against the mortar, you will make a concave joint—a common joint but one of the best. Tooling not only affects appearance but it makes the joint watertight, which is the most important function. It helps to compact and fill voids in the mortar. Fig. 33 shows concave joint tooling.

CHAPTER 12

Concrete Anchors

Whenever possible, the need to fasten to concrete should be anticipated and bolts set into place at the time of pouring. An example of this is shown in Fig. 10 in Chapter 6. Held in place with cross-pieces, the concrete holds the anchor bolts firmly in place when cured.

Fig. 1. Showing one of the ½" bolts cast-in-place when the foundation was poured.

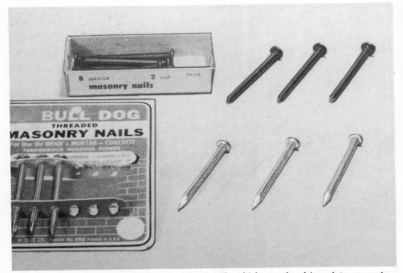

Fig. 2. Examples of tempered, carbon-steel nails which may be driven into concrete.

Fig. 3. Mortar nails which are tapered and have four sides. They can be driven into mortar.

Cast-in-place bolts are used to fasten $2'' \times 4''$ or $2'' \times 6''$ sills to the foundation of a frame home. They are usually ½-12 bolts (½ " in diameter and 12 threads to the inch) to take a flat washer and nut. Fig. 1 shows the bolt in place.

Even though concrete is hard when cured, it is still possible to fasten to concrete. Special fasteners for concrete are used, and special tools are needed to install the fasteners. These special items consist of steel nails, sleeves which hold wood-type screws, and anchors which hold machine-type screws. The holes for the sleeves and anchors are made with special bits.

CONCRETE NAILS

The nails shown in Fig. 2 may be driven into concrete. They are made of tempered carbon steel and have thread-like serrations along their lengths. The tapered nail in Fig. 3 is for driving into the mortar between bricks and concrete blocks.

Driving a concrete nail into concrete is not as easy as driving on ordinary nail into wood. It must be driven straight and with no wobble. Otherwise the hole may be too large to grip the nail. Use a 4-lb. sledge rather than a carpenter's hammer. Tapered nails

Fig. 4. A tool designed to drive special nails and bolts into concrete.

drive into mortar rather easily. Some types of mortar contain air-entrained cement in their mix. Tiny air bubbles in the mortar give way to the nail when it is driven in.

The need for driving nails into concrete perfectly straight has brought out special devices to do this, as well as special nails and screws for use with these devices. *Shure-Set* and *Ramset* are two names of concrete fastener tools and equipment. One set is shown in Fig. 4. The devices that may be driven into concrete include long nails, short nails, and threaded screws. The four illustrations in Fig. 5 show the use of this tool for fastening the back of a workbench to a concrete basement wall.

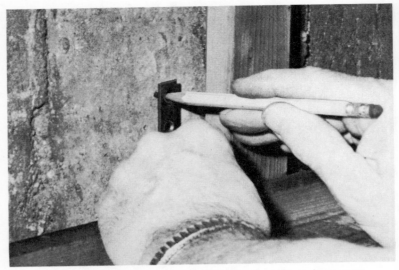

(A) Marking the spot where you wish to drive the nail into the concrete.

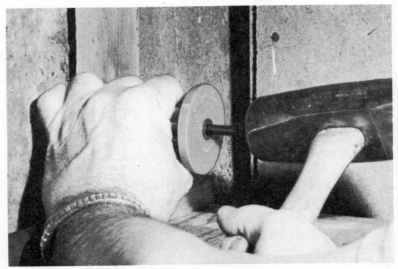

(B) A small sledge hammer is used on the tool driver.

Fig. 5. Steps in fastening with

(C) The stud held securely in the concrete.

(D) A bracket secures the back of a workbench or shelf.

the tool is shown in Fig. 4.

TYPES OF ANCHORS

All mortar- and concrete-holding devices (except one) are similar. They are made of soft material, have an inside thread to take the screw (whatever kind it is), and expand against the walls of the hole for their holding power as the screw is turned in. A screw cannot hold directly in concrete or mortar, but must be held by an anchor.

For heavy-duty bolting, the anchors are made of soft metal, usually lead. For lighter loads, the popular plastic anchors, made specifically for plaster walls, may also be used in concrete. They do have one advantage; they are rust-proof, but do not try to use plastic if the load indicates a sturdier type.

Fig. 6. Various types of lead anchors for wood, metal, or lag screws.

Most anchors are made for use with standard wood screws ranging in size from the tiny #5 to the giant #24 size. One type is made for use with lag screws, which are heavy-duty screws with threads and a taper like wood screws (but much coarser) and with a square head which requires turning with a wrench. Another type accommodates a carriage bolt, which is a heavy-duty, nontapered, coarse screw with a round head. Another type is threaded for machine screws—the smaller-sized screws with screwdriver-slotted heads. Fig. 6 illustrates the most popular lead anchor for wood screws. It comes in sizes to take wood screws as small as #6 only ¾″ long, up to sizes to take a 2″ #24 wood screw. These anchors are also made with a fiber sleeve which gives them greater holding power.

Fig. 7 shows an anchor for lag bolts. These anchors are available in many sizes, accommodating a ¼″ to a ¾″ lag bolt. One type anchor is also shown in Fig. 6. The largest size is capable of handling loads as high as 4000 lbs. A machine-screw anchor is shown in Fig. 8. These are sometimes known as *Ackerman-Johnsons* after the name of the original manufacturer. Different sizes are shown in Fig. 9. The machine-screw anchor is somewhat different from those mentioned above. It is made of two parts—a hard metal inner core threaded on the inside for the machine screw and tapered at one end, and an outer sleeve of lead. These anchors are fastened into the hole with a driver tool (shown in Fig. 9). The anchor is placed in the hole, tapered end inserted first, and the driver is placed against the other end. Hammering on the end of the driver forces the outer lead sleeve in and sideways against the walls of the hole.

The machine-screw anchor holds more securely than any of the others. They are made for machine screws as small as 6-32 and as large as 1″ in diameter, and are capable of holding up to 12,000

Fig. 7. Lag screw expanding a lead anchor.

Fig. 8. A typical machine-screw anchor.

lbs. The disadvantage of this type is the need for a driving tool the same size as the individual anchor size.

DRILLING THE HOLE

Two methods are used to put a hole into concrete or mortar. One method uses an electric drill with a special carbide-tipped bit (A in Fig. 10). The other method uses what is called a "star drill." It is not a drill at all, but a tool that chips out the concrete as you pound it with a heavy hammer (B in Fig. 10). Both of these come in different sizes for making different diameter holes to accommodate various sizes of anchors.

The carbide-tipped drill bits (also called masonry drills) come in ¾₆″ to ¾″ sizes. Those up to ½″ have ¼″ shanks to fit the popular ¼″ electric-drill chucks. A star drill has a flared four-bladed end and is rotated as you pound on the end of it with a small sledge hammer.

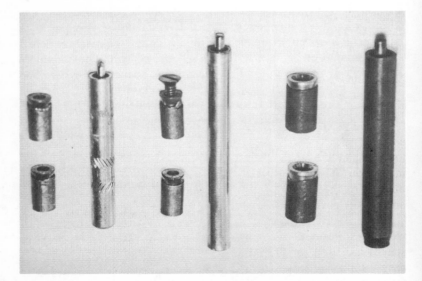

Fig. 9. Various sizes and types of machine-screw anchors, and the tool used to insert them.

Fig. 10. Special bits for making holes in concrete. (A) A carbide-tipped drill bit. (B) A star drill used in conjunction with a hammer.

The four forward edges chip away at the concrete. Wear a heavy work glove on the hand that holds the drill. Otherwise, your hand can become sore holding and turning the drill. In addition, a glove is good protection in case you slip with the hammer.

After you have made a hole in the concrete or mortar, clean out all remaining dust. It is best to use a short piece of small-diameter rubber or plastic hose. Drilling holes in brick is not recommended. It is always easier to go into the mortar between the bricks, and it can be as secure as you will ever need. Holes must be as deep as the anchor or as deep as the full length of the screw if it is longer than the anchor. You can tell which is longer by visual inspection. The diameter of the hole must be a snug fit for the anchor, or the anchor will turn in the hole as you turn the screw, and it will never grip the sides.

Star drills come in sizes from ¼″ to ¾″. A carbide-tipped drill works best in one of the newer variable-speed electric drills. Slower speeds are used for the larger-sized drills. Drilling into mortar is easy—almost as easy as drilling in wood with an ordinary drill. Concrete, however, is much harder. Effective and fast drilling in concrete requires a great deal of pressure on the work.

CHAPTER 13

Repairing and Patching

To patch is to fill cracks or surfaces areas in concrete that have broken as a result of settling, the ravages of weather, or other abuse. Cracks will occur in walks, driveways, patios, and in garage and basement floors. Large cracks are usually the result of heaving or of voids developing in the earth base under the concrete slab. They can also result, of course, from carrying loads heavier than the concrete thickness is designed to tolerate, and (as explained earlier in Chapter 5) by improper reinforcement.

Cracks should not go unmended, no matter how small. If allowed to remain, the crack will become worse with time. In cold weather, the edges will be broken away each time snow melts into the cracks and refreezes.

WINTER EFFECTS

Heaving is a frequent cause of damage in the cold climates of the northern half of the country. The earth under the slab will settle a little, thus creating voids. Water may seep into the voids following a rain just before the temperature drops below the freezing point. In winter, the water is subject to alternate freezing and thawing. When water freezes it expands, and it can exert a force great enough to press up from the underside of a slab and crack it.

EROSION

Voids may develop under the slab as a result of erosion of the earth base; a leaky water pipe or drain tile passing beneath may be the cause. Cracks from erosion frequently occur at the corners of the concrete.

HAIRLINE CRACKS

The repair of cracks depends on how extensive the damage is. For hairline cracks, a "grout" made of portland cement and water is best. Put some portland cement in a coffee can that has been wiped clean of coffee dust, and add a little water; just enough water to make a thick paste. Use a putty knife or pointed trowel to force the grout into the crack, and smooth it off level with the rest of the slab. To assure good adherence of the grout to the concrete, moisten the old concrete with water for several hours before you add the grout. If the concrete has absored some water before the repair, it will not draw moisture out of the grout. After wetting thoroughly, brush out excess water and dirt from the crack and apply the grout.

Even fine-crack repairs must be properly cured. After a couple of hours, when the patch has taken a set, cover the area with a piece of plastic sheeting or a board. For the next five days, lift the covering each day and sprinkle the repair with water. Unless the repair is properly cured, the same as in the case of new concrete, it will be weak and eventually break out again.

LARGER CRACKS

Large cracks will call for more work. If the crack is over $1/16''$ in width, it should be chipped away and widened with a cold chisel. But more than that, the crack must be provided with either straight or undercut sides to provide a better bond of the new concrete to the old cement.

Fill the larger cracks with a mixture of cement and sand. Use a mixture of about 1 part cement to 3 parts sand. Thoroughly mix the cement and dry sand, then add water until it is a stiff mixture that can be worked with a trowel.

231

To assure good adherence of the new concrete to the old, mix a little portland cement with water to the consistency of paint and "paint" the edges of the cracks in the old concrete. However, be prepared with the new matching concrete mix to apply it to the crack almost immediately. The cement painting must not be allowed to dry before completing the patch.

All repairs of concrete must be preceded by a soaking of the old concrete with water—the longer the soaking, the better the repair will be. However, when fresh concrete is applied to old work, there must be no surplus of water to dilute the new concrete. A good time for repairing concrete is soon after a long rain.

A good way to moisten the old concrete is to let water run from a garden hose onto the cracked area overnight. You will only need to turn the water on a little—just enough for a trickle of water to cover the broken area. Important to a good repair is for the concrete around the crack to be perfectly clean and damp.

The best quality patching cements are latex and epoxy types. These adhere to old concrete more securely than standard mixtures of concrete. However, they are too costly for any extensive jobs which require a lot of material.

SURFACE BLEMISHES

After a year or so, sidewalks and driveways, may have thin slices broken off of the surface, or fine cracks or dust-like breakup of the top surface. These blemishes are primarily the result of incorrect handling of the concrete at the time it was poured and finished; they are defects that could, and should, have been avoided in the first place. The following are the principal defects, and their causes.

Scaling

Scaling is the condition where thin slices, or pieces, flake off the top surface of the concrete, usually to a depth of from 1/16″ to 3/16″. There are three ways to prevent this.

When early cycles of winter freezing and thawing of the surface occurs on newly poured concrete, it is because the concrete was not warm enough when poured and was not kept warm for at least 5 days. Concrete must be poured and maintained at a temperature of not less than 50°F until cured. When flaking occurs some time after curing due to freezing and thawing and from the use of deicing salts, it is because air-entrained concrete was not used. Cold climate areas should use air-entrained concrete to protect against the ravages of freezing action. Scaling can also be the fault of poor workmanship. Finishing concrete (striking, floating, etc.) before the free water surface has evaporated is a common fault. When you work the top surface while free water is still standing, the water works itself down below the surface and causes some separation of the cement from the sand just under the surface. Never do any finishing operation to concrete while free water is present. If the water seems to be taking forever to evaporate, you can help it along by setting a fan to blow across it, or even draw a hose across to remove the deepest part.

Crazing

Crazing appears as a number of fine cracks in the surface of newly cured concrete and looks like the cracks of crushed eggshells. The usual cause of this is too rapid drying of the surface during the curing process. The prevention, of course, is proper covering of the concrete and the application of water—especially in climates of low humidity and in the presence of high winds. Best curing is obtained when the entire slab cures evenly and slowly. Crazing can also be the result of premature finishing before all the surface water is gone—the same as for scaling.

Dusting

When the surface of concrete breaks away in fine powder form, this is called dusting. If there is an excess of clay or silt in the mix, these tend to rise to the surface and result in a weak concrete at the top. The prevention is to use clean and well-graded sand and aggregate. Again, premature finishing can result in weak concrete at the top that can wear off as a dust.

REPAIRING SURFACE DAMAGES

Besides poor original workmanship, the ravages of weather or hard use may result in crazing and dusting. These are surface

233

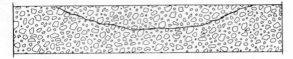

(A) Incorrectly installed patch. Feathered edges will break down under traffic or will weather off.

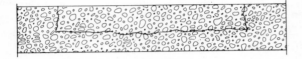

(B) Correctly installed patch. The chipped area should be at least 1″ deep with edges at right angles to the surface.

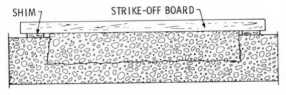

(C) Correct method of screeding a patch.

Fig. 1. Installing a patch.

damages that can become progressively worse with time if they are not repaired.

If the damaged area is large and deep, you may need to cut out the piece and replace it. It is usually not necessary to cut it completely out (down to the soil); only cut it out down to good concrete.

Use a cold chisel and a short-handled sledge hammer, or a chisel-pointed power hammer. Chip out the damaged area to where you have good solid concrete, with straight or undercut sides in the cutout area. Fig. 1 shows how this is done. Again, soak the concrete overnight and hose out all loose dirt and concrete particles. Use an old towel to pick up the excess moisture, if necessary.

Coat the bottom surface and sides of the area with a concrete paint (as mentioned for the wider cracks), and apply new con-

234

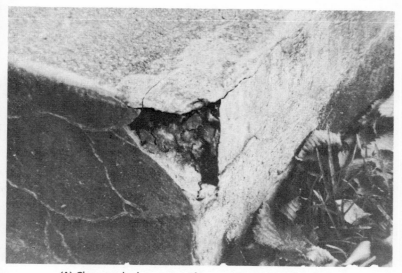

(A) Clean out broken corner of a concrete step and dampen it.

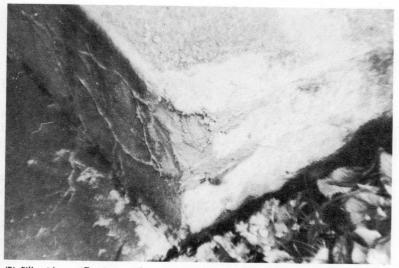

(B) Fill with a stiff mixture of concrete and sand. The sides can be smoothed after the concrete has taken a slight set.

Fig. 2. Repair of a broken step corner.

235

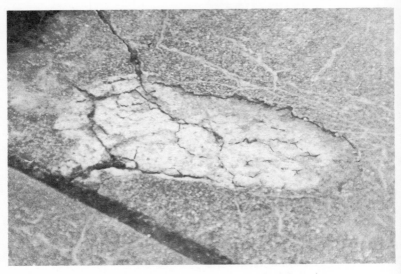

(A) The old pieces of broken concrete can be left in place.

(B) Use a steel-bristled brush to remove small pieces.

Fig. 3. Step-by-step repair of a shallow surface

(C) Sweep out all dust and loose pieces; wash out with a garden hose.

(D) Remove lid from a 5-lb. can of latex cement. The liquid latex is inside in a separate can.

break in a concrete sidewalk.

(E) Pour the liquid latex into the cement in the large can.

(F) Stir thoroughly; the results will look like a heavy paste.

Fig. 3. Step-by-step repair of a shallow surface

crete. If the patch is deep, use a standard concrete mix (including larger aggregate) for the sake of economy. If it is shallow, use a sand-and-cement mix. Apply the new concrete in ⅛" or ¼" layers, smooth each layer out with a trowel, and allow the shiny surface moisture to evaporate before applying the next. When the top layer is on, level it with a strike board, and trowel or float when all excess moisture is gone. Try to match the texture of the old concrete. If it has a slightly rough texture (as is usual for walks) use a wood float; if it is smooth, as in the case of many patios, use a metal float.

REPAIRING BROKEN CORNERS

If a crack has occurred near the corner of a horizontal slab of concrete or at the corner of a step, a piece will frequently be broken off. You have your choice of two methods for repair. You

(G) Apply the latex concrete to the break and trowel smooth, feathering out the edges. After the concrete has set, cover it and keep moist for at last 5 days.

break in a concrete sidewalk.

may re-cement the old pieces back on, or, in the case of broken sections of a horizontal slab of concrete, leave the pieces there and add new concrete to the top.

Cementing the old piece back on is recommended in the case of a corner of a step if it has come off in one piece. After all, the old piece is the proper size, shape, and color. A dependable job of cementing on a broken piece calls for the use of latex or epoxy concrete. Standard portland cement concretes just do not have the power to bond to old concrete with sufficient strength; latex and epoxy concretes do.

Again, thorough cleaning and dampening of the old concrete is required. Then "butter" the broken piece with the new concrete, and hold it in place. If the piece is off of a vertical section of concrete, you will need to place a board or something against the repaired piece until the concrete has stiffened enough to be self-supporting. There is plenty of time after this stiffening to clean off the excess concrete that oozes out of the joints between the pieces. Finish off the crack the same as you would a hairline crack.

If the corner has crumbled off into several pieces, repair can be made by filling in the corner with fresh concrete (Fig. 2). In a walk or other horizontal section of concrete with broken pieces, the pieces may be left in place and concrete poured over them to the level of the rest of the slab. A thorough cleaning and dampening job is required first. Let that section soak overnight with water from a slowly running hose; then sweep out the excess water and all dirt. Paint the tops of the broken pieces and the edges of the slab with a cement paint, and, before it dries, pour in the concrete.

The concrete may be 1 part portland cement to 3 parts sand or a *Sakrete* sand mix. Latex or epoxy concretes will give you a stronger and longer lasting job, but be sure to check on the economy first. These latter two are a bit expensive for large repair jobs.

USING LATEX CONCRETE

Features of latex and epoxy concretes are their power to adhere to old concrete and their ability to be feathered out to very thin depths. When using these concretes it is not necessary to chisel out vertical sides on the old concrete. Use a stiff wire brush and clean out as much of the old loose bits and pieces as possible. Dampen the area as for any concrete repair. Apply latex or epoxy concrete in layers to the open area, building it up to the level of the rest of the slab. Fig. 3 shows a step-by-step pictorial story of how to repair a shallow surface break with latex concrete.

Fig. 4. Latex or epoxy concrete is excellent for repairing cracks in a wall.

In Fig. 4, latex concrete is being used to fill cracks in a concrete wall.

Latex concrete may also be purchased in bags as a dry ingredient. Pour out only as much as you need for the job, add a little water, and you have a tough concrete that is ideal for small jobs. Seal and store the rest of the bag in a dry place.

REPAIRING DAMP OR LEAKY BASEMENTS

Of all the problems involving concrete, one of the most vexing is the accumulation of moisture on the walls or floor of fully concreted basements. There are several causes for the accumulation of moisture; some are easily cured, and some are not so easy. Most problems are the result of original construction faults.

Condensation

Walls may become moist through condensation, which is not the fault of workmanship during construction. If the air in a basement is moist because of high summer humidity or an imperfectly vented clothes drier, and the concrete walls of the basement are cold, the moisture of the air will condense on the walls. This is the same sort of condition that causes cold water pipes to "sweat." The problem is usually a summertime ailment.

There is only one cure—reduce the moisture content of the air. Check the clothes drier to make sure it is vented to the outside. Keep a fan going to circulate the air; this will not get rid of the moisture, but it will evaporate the condensation off of the walls. The best answer, if the source of the moisture cannot be corrected, is to use a dehumidifier in the summer months.

To be sure the problem is only that of condensation, make this simple test. Fasten a bright tin sheet about 6″ square to the wall. Use adhesive tape on the corners to hold it in place. Bright aluminum or any other bright metal will also do. Leave it there for about an hour, and then inspect the surface. If it is moist, the problem is condensation; if it is dry, the problem is seepage.

Seepage

Seepage is the slow leakage of water through fissures in the walls or where the walls join the floor. It is most evident after a steady rainfall or an over watering of the lawn outside.

Seepage, in most cases, is the result of poor construction. It can be due to:

1. A poor concrete mix at the time the walls were poured by the contractor, resulting in a porosity that allows water to come through.

2. Inadequate seal between the walls and the floor where they join in the lower corners.
3. A poor job of coating the outside of the walls with a bituminous coating.
4. Inadequate installation of drain tile on the outside.
5. Failure to slope the subsoil away from the house.

Sometimes the drain tiles, or the outlets from them, will become clogged with tree roots which prevents the water from flowing away fast enough. A possible contributing factor to excess water in the ground can be the downspouts from the roof gutters. If they do not connect to underground pipes for disposal through the sewer system, they should have good long leads to take the water several feet away from the edge of the house.

Outside Drainage

The best correction for seepage is from the outside—the earth side of the wall. It generally means digging trenches all around the house, and down to the depth of the walls. The standard treatment is to chip away identifiable cracks in the wall and fill them with concrete. Next, the walls must be given a treatment of hot bituminous material. Two coats should be applied—one with brush strokes in one direction, and the other with brush strokes in the direction perpendicular to the first. Use either a coal-tar pitch meeting builders standard ASTM D-450, or asphalt meeting ASTM D-449 standards. Both should be type B. The walls must be scrubbed clean before the tar is applied.

Install drain tile around the house with a slight slope so water will run down into the storm-sewer intake. A deep layer of gravel or stones averaging 1" to 1½" in diameter should be placed over the tile.

Inside Wall Treatment

Seepage that occurs only seasonally or occasionally can be corrected from the inside with a couple of coats of concrete or cementitious paint. Fissures in the concrete walls can be covered by applying additional concrete in the form of two layers (each about ¼" thick) to the inside. This new coating will protect against

243

leaks up to about ⅛″ in size, but not larger. If there are any holes or cracks larger than ⅛″, these must be filled first before applying the new coating.

CONCRETE COATING

Make a concrete coating from portland cement or masonry cement. Mix 1 part cement with 2½ parts of damp, loose, mortar sand. Add water to make up into a stiff plaster that will stick to the walls.

Fig. 5. Recoating a concrete basement wall.

Scrub the walls down with a stiff brush and water (remove all loose dirt or dust), then hose the walls. The walls should be damp when the concrete coating is put on. Apply about a ¼″ thick coating of the plaster. When it has stiffened a little, go over it with a wire brush to put on scratch marks. Allow it to cure for 24 hours, then apply a second coat (Fig. 5). After it has set hard, spray it with water about twice a day for at least 48 hours to keep it damp and thus allow hard curing.

CEMENTITIOUS PAINT

When seepage is of little consequence, it may be sufficient to apply one of the special paints which are sold as waterproofing paint. They are available in several colors. They need only to be scrubbed into the surface of the concrete.

When applying any coating to the walls to prevent seepage, the walls must have stopped leaking. While walls are dampened before the application of a new coating, they must not be wet. Also, there must not be continuous seepage while the coating is curing. Wait for a dry season when there is no seepage, or if it is the result of openings larger than ⅛″, fill them first.

WATER UNDER PRESSURE

Leakage can result from water being under pressure and coming through cracks or holes. This can occur after a heavy downpour of rain that has lasted for quite a while. The sewer system cannot carry off the water fast enough, so it will back up. This is the reason for standpipes and screw plugs in basement drains. But even with the drain properly stopped, water may be forced through tiny cracks in the concrete because of the pressure built up outside. Sewer tiles are not watertight at the joints, and water under pressure can leak out and develop pressure under the concrete floor of the basement.

The most often-guilty culprits for leaks are the joints between the walls and the floor. When this happens there is a hissing sound made by the water coming through under pressure. Obviously, the correction is to seal off the basement to perfection.

HOLES AND LARGE CRACKS

Holes or cracks greater than ⅛ ″ must be patched with concrete. Patching concrete may be the same as described previously. Either mix concrete by using 1 part portland cement to about 3 parts sand and adding water to make a stiff mortar, use the *ready-mix* packages, or use latex or epoxy concrete. The latter two hold to old concrete more tenaciously, but are costly by comparison. The decision as to which to use depends largely on the amount of work to be done. If water is seeping or running through at all times, use a fast-setting concrete especially made for the purpose.

Whichever concrete is used, an important part of the job is preparing the cracks or openings for patching. This is important. New concrete (except epoxy and latex concretes) does not readily adhere to old concrete. The answer is to chisel out an inverted V-shaped notch. Using a cold chisel, cut out all loose concrete, and cut the notch so the inside of the opening is wider than the outside. Fig. 6 shows the right way and the wrong way. The front of the opening need not be any larger than the original size of the crack, but it is almost impossible to keep from taking some of the edge off during chiseling. However, there is no harm in enlarging the crack a little; the important thing is to under-cut to keep the new concrete in.

After chiseling out the cut, clean out all loose concrete or other loose particles. Flush out the cut with a stream of water from a hose. Thoroughly mix 1 part cement and 3 parts sand. Add only enough water to make a thick mortar which is capable of standing up under its own weight. After first moistening the inside of the crack—push the concrete into the crack with a pointed trowel or putty knife. Push the concrete in well; make sure there are no air voids. Finish off the outside flush with the wall.

If the water is running in at all times, you will not be able to stop it with regular concrete. In this case, prepare the cut as mentioned above and place a small pipe in the opening at the bottom of the cut. Water will come out through that pipe while you continue with regular patching above it. Patch up to the pipe.

After the concrete has set for a couple of hours, but before it has hardened too much, remove the pipe. Prepare a plug of special fast-setting concrete (frequently called "wall plug") by rolling it in your hands. Then, quickly push the plug into the hole at the bottom. The special fast-setting plug will begin to harden in a couple of minutes.

A most annoying point of seepage is at the edges of the floor where the floor meets the walls. Most concrete floors and walls are poured separately when the house is built. Frequently there is not a perfect seal between the two. If the problem is corner seepage, chisel a 1″ to 1½″ groove all around the floor where it joins the walls. Begin with a line about 1″ above the floor on the walls, and cut diagonally down into the joint. Pour hot tar or tar mixed with sand into the grooves. A better but more expensive seal may be made by using latex or epoxy concrete. Mix either according to directions and push it well into the groove after thoroughly flushing out all loose particles.

The groove surfaces must be completely dry and free of any moisture before filling with tar or tar/sand mix. On the other hand, filling the grooves with any concrete-type filler calls for moist (but not wet) surfaces to prevent fast absorption of the moisture from the concrete patching material.

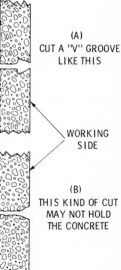

(A) CUT A "V" GROOVE LIKE THIS

(A) Right way—cut a V-groove into the concrete.

WORKING SIDE

(B) THIS KIND OF CUT MAY NOT HOLD THE CONCRETE

(B) Wrong way—this kind of cut may not hold concrete.

Fig. 6. The right and wrong way to chisel a crack for filling with concrete.

247

CURING

It cannot be emphasized too strongly that the curing of concrete repairs is just as important as the curing of poured concrete on new projects. Unless the repair is kept moist and prevented from drying too fast, the concrete patch will shrink and develop new cracks, or break out altogether. After a repair is set enough to carry the weight of a cover, use any of the materials mentioned in Chapter 8, and keep it moist by occasional watering. Keep this up for 5 days, after which you may remove the covering and the repair will be as strong as the original concrete with no need for additional work for many years.

CONCRETE PAINT

The demand for color and the growing acceptance of concrete as a decorative material and surface finish have resulted in an increasing use of paints on concrete. In the following discussion several broad groupings are made of the many different kinds of paint available.

The selection of paint for a particular case will depend on:

1. The condition of the concrete.
2. The service to which it will be subjected.
3. The effect desired.

Portland Cement Paint

This paint consists largely of portland cement, especially prepared and ground with other materials. Limeproof and sunproof mineral pigments are ground with the other materials to produce the colored product. A wide range of colors can be had, the possibilities limitless if the standard available colors are blended. It is sold in dry powder form to be mixed with water before applying.

Portland cement paint should be distinguished from ordinary cold-water paints. Sometimes these products are referred to as cement paints but are not true portland cement paints. Many of them contain large proportions of hydrated lime, casein glues, or other materials. Some of these materials offer little resistance to weather or other adverse conditions and can serve only as temporary color coatings. For outdoor and locations subject to moisture, only paints which the manufacturer can definitely guarantee to withstand all conditions of weather and moisture should be used.

A true portland cement paint is easily applied and bonds with any concrete surface, stucco, concrete masonry, common brick, soft tile, limestone, or any other type of cement work or masonry which presents a clean surface having some absorption. It is not recommended for application on wood or metal, enameled brick, vitrified or glazed brick or tile, surfaces saturated with oil or grease, nor on surfaces previously coated with oil paints unless such paint is completely removed. It is not recommended for concrete floors, except the floors of swimming pools, wading pools, ponds, and shower rooms.

An outstanding advantage of portland cement paint is that it can be applied successfuly to masonry surfaces that are damp. In fact, all surfaces must be uniformly damp when the paint is applied. Portland cement paint can be applied to concrete immediately after the forms have been removed, or to fresh cement plaster or stucco.

Application—Portland cement paint should be mixed and applied in accordance with the manufacturer's directions. Dirt, oil, grease, or efflorescence should be thoroughly removed from the surface to be painted. Oil and grease can be removed by scrubbing with gasoline or naphtha. Efflorescence can usually be removed by scrubbing with a 10% solution of muriatic acid. This should be followed by thoroughly rinsing with water.

As previously stated, it is essential that the surface be uniformly damp, not wet, when the paint is applied to avoid too much suction and to aid in curing the thin paint coat. A garden hose provided with a fog nozzle will be satisfactory for dampening the walls and curing the paint coats.

A brush with relatively short, stiff fiber bristles is recommended. Ordinary paint or calcimine brushes are not satisfactory. The stiff fiber brush, using a scrubbing motion, is required to force the paint into the open texture of the surface and the mortar joints.

After the first coat of paint has hardened sufficiently to prevent injury to the surface, it should be cured by keeping the surface damp for at least 24 hours. This can be done by spraying at required intervals. The surface should be wetted down again just

249

before applying the second coat. At least 24 hours should elapse between the coats. The second coat should be moistened with a fine spray as soon as the paint has hardened sufficiently to prevent marring the surface. It should then be kept damp as long as practicable, but not more than two days.

During warm, hot, and windy weather the paint film will harden or "set" more rapidly than during cool or cold weather. Therefore, the paint will need to be sprayed sooner during warm weather than during cold weather.

Organic Paints

Practically all other paints are largely organic. It is difficult to secure a good job with these paints on concrete where moisture may enter from behind. They form impermeable films which may be pushed off by moisture in the concrete.

While little preparation of the surface is necessary for the application of portland cement paints, the organic paints require considerable care and attention. The concrete must be perfectly dry and well seasoned. Oil, grease, and efflorescence should be removed. Old concrete should be wire brushed to remove loose particles. New concrete work should be given at least 8 to 10 weeks to dry after the moist-curing period before the paint is applied.

When the concrete has dried thoroughly following the curing period, it should be given a neutralizing wash to prevent saponification of the oils. A solution consisting of 2 to 3 lbs. of zinc sulfate crystals per gallon of water, or a solution containing 2 lbs. magnesium fluosilicate per gallon of water may be used. A small amount of pigment may be added to the wash so it will be seen when applied, preventing the skipping of areas. At least 48 hours should be allowed for the solution to dry. When thoroughly dry all protruding crystal should be removed by brushing. This treatment is not necessary on old concrete or old stucco.

After the neutralizing wash has dried, the concrete surface should be given a binding and suction-killing treatment, consisting of one or more coats of oil or varnish carrying some pigment. A suitable primer may be made by mixing one gallon of oil paint, one-half gallon of chinawood-oil spar varnish, and one quart of turpentine or similar thinner. Special chinawood-oil priming paints for concrete are also available. The prime coats should be allowed to dry thoroughly, after which any of the organic paints can be applied in accordance with the manufacturer's directions.

Cement and Oil Paints

Cement and oil paints is the general term used to include all paints made by grinding portland cement and pigments in an oil vehicle. The vehicle may be linseed oil, a mixture of linseed oil with chinawood and other oils, or *Cumar* in a hydrocarbon solvent.

These paints dry to a hard, flat finish somewhat rough to the touch. They adhere tenaciously to concrete and other masonry, but are not recommended for floors. Fine silica sand is sometimes mixed with the final coat.

Oil Paints

There is a wide variety of paints and enamels made by grinding white lead in a vehicle such as linseed oil, and adding other pigments to produce color. Modern paints are also made from other pigments; for example, aluminum or zinc, and from oils other than linseed.

In some cases the pure metals instead of the oxides are used. Such paints made by reputable manufacturers are suitable for use on concrete, provided they are applied over properly treated surfaces.

Special and Proprietary Paints

Many of the newer paints and lacquers consist of resins, gums, synthetic plastics, and a wide range of other materials dissolved in hydrocarbon or other suitable solvents. These paints should be used in accordance with the directions of the manufacturer.

Painting Concrete Floors

A satisfactory paint job on a floor of any type presents a difficult problem, and in this respect concrete is no exception. Trucking and dragging of boxes, crates or other loads, and heavy foot traffic concentrated in small areas are destructive to paints on any surface. Where traffic is light or moderate, and evenly distributed, paint films on floors will give good service. It should be appreciated that repainting will be necessary at intervals depending on the service requirements.

Most manufacturers of lead and oil paints make special floor paints in which they embody abrasion-resisting pigments. These are suitable for concrete floors when carefully applied. Paints with a phenol resin base have also given good service, as have a number of paints prepared especially for use on concrete floors. Some of the special floor paints contain rubber, giving a hard, durable surface coating resistant to water and to the alkali in concrete.

Application—It is essential to have the floor perfectly dry when paint is applied. Basement floors, where moisture penetrates the slab from below, should not be painted because the moisture will cause the paint film to peel and chip off. Paint on some basement floors has been unsatisfactory because warm, moisture-laden air has condensed on the floor slab before the paint dried.

Painting should be done when the atmosphere is dry. New concrete should be thoroughly seasoned and neutralized, as previously stated, before paint is applied. The magnesium fluosilicate treatment is recommended since it hardens the floor surface at the same time. Any undue roughness of the surface should be removed with a carborundum brick or coarse emery cloth. The surface should be thoroughly cleaned, removing oil and grease by scrubbing with gasoline, naphtha, or special solvents, allowing them to evaporate before painting.

While most manufacturers recommend two coats on concrete floors, experience indicates that three coats give results which more than compensate for the extra cost. Manufacturers' directions for application should be followed.

Epoxy Enamel

Like epoxy patching concrete, epoxy enamel is a two-solution liquid which, when mixed together, becomes a very hard glossy enamel with tremendous durability. It adheres well to any surface, including wood, porcelain, brick, and concrete. It is available in colors or white, plus pigments that may be added.

For good adherence, the concrete must have a slightly roughened surface. This is done by etching with muriatic acid. First scrub the concrete with trisodium phosfate (TSP) to remove dirt and grease. Then apply diluted muriatic acid with a sponge.

TSP is injurious to growing plants and lawns. After the concrete is cleaned, use a garden hose to thoroughly rinse off all TSP, and heavily dilute that which may run off onto a lawn or garden area.

Muriatic acid is harmful to the skin and to clothing. Wear rubber gloves and use a sponge when applying it. Dry crystals of muriatic acid are easier to handle. After applying it, rinse off thoroughly with running water, and allow the concrete to dry completely before painting.

The two solutions that make up epoxy enamel are mixed well, then allowed to stand for about a half hour for curing. Mixed enamel not used within about six hours must be discarded, as it becomes too stiff to apply. The enamel must be allowed to "age" for from 5 to 7 days before the concrete is to bear any traffic.

CLEANING CONCRETE

Concrete may be cleaned with either of two solutions. Use TSP, as previously mentioned, or aromatic hydrocarbons. Dilute according to the instructions with the package. The amount of dilution varies with the concentration of the chemical purchased. Grease, as on garage floors, and other stains are easily removed. Flush well with running water.

CHAPTER 14

Projects with Concrete

On the basis of the information contained in the previous chapters, any contractor or individual should be able to perform any operation with concrete except very large projects requiring engineering. This chapter describes a few simple, around-the-house projects which can be done by apprentice masons, or even the homeowner.

Perhaps the easiest of projects are those for slab-type concrete, such as for sidewalks, driveways, or patios. Such projects, when

Fig. 1. Clear the area and place a grid of 1 x 4 redwood boards to make 25- or 30-inch squares.

small, can be reduced in complexity when using prepared mixes in which all the necessary ingredients are contained in the proper proportion in one sack. All that is necessary is to add water as directed on the package. Prepared packages are available in several types of mixes. Some have cement, sand, and aggregate for general concrete use. Others have cement and sand only for small patching jobs. Some have mortar mixes. They can be purchased in a number of sizes, up to a 90-lb. sack. Local suppliers may package them under their own name. The best known of the national brands is *Sakrete*.

A convenient table for mixing ingredients for small jobs is shown in Table 1. The ingredients for concrete volumes from 1 cu. ft. to 1 cu. yd. are shown. Power mixing is recommended; note the total weight to be handled by hand mixing for even only ¼ cu. yd. of concrete.

Table 2 gives information in simplified form of total cubic yards of concrete needed for various size areas of slab construction, and for various depths. The usual depth of small slab jobs of the type mentioned above is 4″.

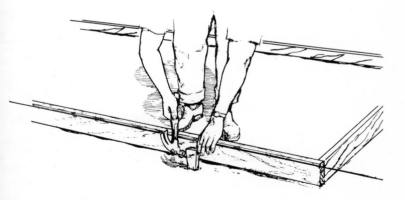

Fig. 2. Use a string guide to ensure level; nail the boards to the stakes set in the ground.

Table 1. Mixing Small Amounts of Concrete

Concrete Required (cu. ft.)	Cement* (lbs.)	Max. Amount of Water to Use (gals.)		Sand (lbs.)	Coarse Aggregate (lbs.)
		U.S.	Imperial		
1	24	1¼	1	52	78
3	71	3¾	3⅛	156	233
5	118	6¼	5¼	260	389
6¾ (¼ cu. yd.)	165	8	6¾	350	525
13½ (½ cu. yd.)	294	16	13½	700	1,050
27 (1 cu. yd.)	588	32	27	1,400	2,100

*U.S. bag of cement weighs 94 lbs. Canadian bag of cement weighs 80 lb.
A 1:2¼:3 mix = 1 part cement to 2¼ parts sand to 3 parts (1-in. max.) aggregate.

Courtesy Portland Cement Association.

Table 2. Cubic Yards of Concrete in Slabs

Area in square feet (length X width)	Thickness in inches				
	4	5	6	8	12
50	0.62	0.77	0.93	1.2	1.9
100	1.2	1.5	1.9	2.5	3.7
200	2.5	3.1	3.7	4.9	7.4
300	3.7	4.6	5.6	7.4	11.1
400	5.0	6.2	7.4	9.9	14.8
500	6.2	7.7	9.3	12.4	18.5

Courtesy Portland Cement Association.

ONE-MAN PATIO PROJECT

The illustrations in Figs. 1 through 11 show the step-by-step procedure of how to make a concrete patio. The job can be done by one man, although not in one day or one week-end. It is intentionally divided into slab sections to permit interruption in the

256

Fig. 3. Cross pieces of the wood grid are notched and nailed solidly in place.

Fig. 4. This is what the finished grid assembly looks like.

work if necessary. Redwood is used for the individual squares and permanently left in place as expansion joints. Redwood withstands the rigors of weather. Although it will change from its fresh reddish color to a gray in time, it does not deteriorate.

This project is further simplified by the use of 90-lb. sacks of *Sakrete*, which is mixed with water one sack at a time in a wheel-

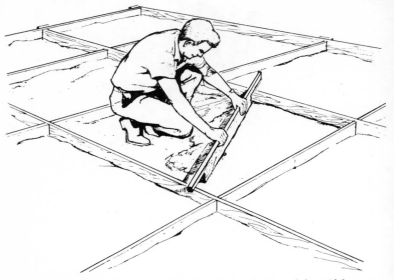

Fig. 5. Pour in sand, and level to about 3" below the tops of the grid frame.

barrow. All illustrations in this series are courtesy of *Sakrete, Inc.* The final step (not shown) is to cover the finished concrete with plastic sheeting for at least five days for curing. For special decorative effects, the fancy finishes described in Chapter 9 could be applied.

A FISHPOND

A fishpond may be simple or complex, depending on the amount of water depth desired, the size of the pond, and the amount of work you are willing to exert. Fig. 12 shows the details for a pond, 6 ft. × 10 ft., which may be built rectangular or bowl-shaped. The rectangular pond holds more water but requires the use of forms for the concrete sides. The bowl-shaped pond has less volume but is easier to build.

The rectangular pond will require very careful excavation to maintain straight sides and bottom. The bowl-shaped is scooped out without too much concern for the exact shape. Excavation must also include facilities for the drain and water supply pipes.

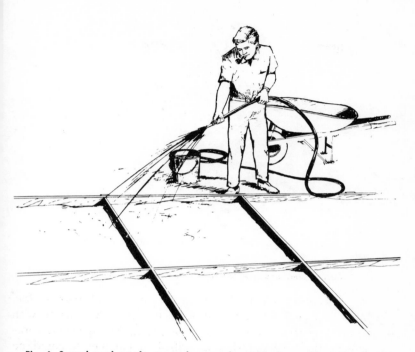

Fig. 6. Spread sand evenly; tamp down, and sprinkle thoroughly with a hose.

After excavation, the next step is to lay in the necessary piping. The outlet pipe should have a threaded coupling at the bottom. This permits removal of the overflow pipe for draining and cleaning the entire pool, when and if necessary. The supply pipe could come up from the bottom, as shown in the center sketch, or from the top side of the pool, as shown in the lower sketch. If a waterfall effect is desired, a system of recirculating pumps and piping must be installed at this time. Many cities forbid the use of a continuous flow of supply water for waterfalls. Furthermore, if the owner is on a water meter, a continuous supply could be very expensive to operate.

Step two is to set reinforcement into place. This could be ⅜″ bars, or 6-gauge screen mesh. The reinforcement will prevent cracking of the concrete in case of water freeze-ups in winter. Use bars,

259

Fig. 7. Dump and mix one sack of concrete mix at one time.

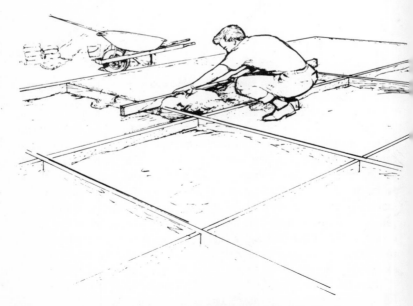

Fig. 8. Pour concrete and screed level with strike board.

260

Fig. 9. Let surface water evaporate; then float. Use wood or steel float, depending on the kind of surface wanted.

Fig. 10. Round corners and put finish on edges with an edger.

Fig. 11. This is what the patio can look like after completed.

chairs, or clean rocks to set the bars or mesh about 3″ above the earth base, and 3″ from the side walls in the case of the rectangular pond. The reinforcement must be held securely so they will not shift during the pouring of the concrete.

For the rectangular pond, step three will be the construction of a four-sided form for the side walls. This must be prepared in advance so it can be lowered into place as soon as the bottom concrete is poured. The form would be 5½ ft. × 9½ ft., for a 6″ thick concrete side wall. It must be firmly braced with cross pieces to prevent bulging inward from the force of the plastic concrete.

With the aid of Table 2, determine the total amount of concrete needed. It will be necessary to prepare for continuous pouring to assure no breaks in the concrete continuity. A power mixer is a must for this project. Standard ingredient proportions as shown in Table 1 could be used for the rectangular pond, but a stiffer mix is needed for the bowl-shaped pond. Concrete with a 1″ slump is about right. Stiffer concrete is achieved by the use of more sand or aggregate, or both.

Screw the overflow pipe in place. Pour concrete into the floor of the pond around the overflow pipe first. Use a straight board in the rectangular pond to level the base concrete as you work away from the overflow pipe. As soon as the surface water has left the area of

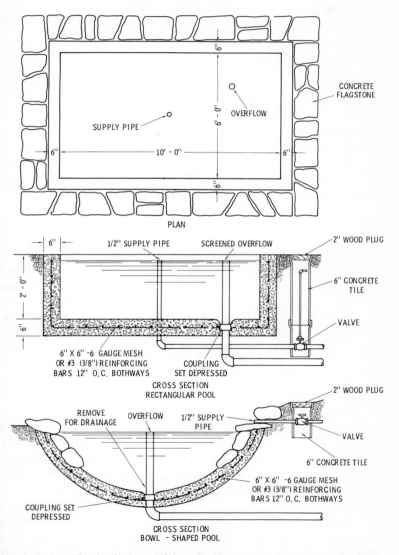

CONCRETE
FLAGSTONE

OVERFLOW

SUPPLY PIPE

6'

6' - 0''

6''

10' - 0''

6''

PLAN

6''

1/2'' SUPPLY PIPE   SCREENED OVERFLOW   2'' WOOD PLUG

6'' CONCRETE
TILE

2' - 0''

VALVE

6''

6'' X 6'' -6 GAUGE MESH
OR #3 (3/8'') REINFORCING
BARS 12'' O.C. BOTHWAYS

COUPLING
SET DEPRESSED

CROSS SECTION
RECTANGULAR POOL

2'' WOOD PLUG

REMOVE
FOR DRAINAGE   OVERFLOW   1/2'' SUPPLY
PIPE

VALVE

6'' CONCRETE TILE

6'' X 6'' -6 GAUGE MESH
OR #3 (3/8'') REINFORCING
BARS 12'' O.C. BOTHWAYS

COUPLING SET
DEPRESSED

CROSS SECTION
BOWL - SHAPED POOL

Fig. 12. Details of a back yard fishpond; either rectangular or bowl shaped.

263

the overflow pipe, use a trowel of displace the concrete around the pipe coupler. Continue pouring concrete into the flat bottom (in the case of the rectangular pond), leveling with a straight board. Use a curved board to screed the bottom of the bowl-shaped pond.

In the case of the rectangular pond, before the concrete has a chance to take a set, place the side forms in place as soon as possible. Immediately pour concrete into the spaces for the sides. For the bowl-shaped pond, screeding and darbying must be continuous as the concrete is poured right up to the top edges. The plastic concrete may have a tendency to flow down the sides but, by continuous screeding, it will hold its shape and soon will set enough to stay firmly in place.

If natural rocks are to be placed along the inner edges, as shown in the lower sketch, do this before the concrete takes a firm set. Indent the rocks into the concrete, and they will cement in place. The surface of the concrete may be floated with a wood or metal float after all excess water has evaporated, or it may be left in rough form.

As shown in the top sketch, the top edges may be made of cut flagstone laid in place, or finished in any other form. Cast-in-place concrete may also be used. Colorful brickwork should also be considered. Curing for a concrete pond is easy. After final finishing, fill the pond with water for about five days. Because of the presence of alkalies in freshly poured concrete, the pond should be filled and drained several times at weekly intervals before fish are put into it.

BARBECUE PIT

The barbecue pit described here is built of concrete block on a concrete slab. It is diagrammed in detail in the sketches of Figs. 13 and 14. The slab thickness will depend on the severity of the weather in the area in which it is built. A 3″ slab is sufficient in the south and along the west coast. A 4″ slab is recommended in other areas where severe freezing may affect the slab. The finished slab should be about 1″ above the ground level.

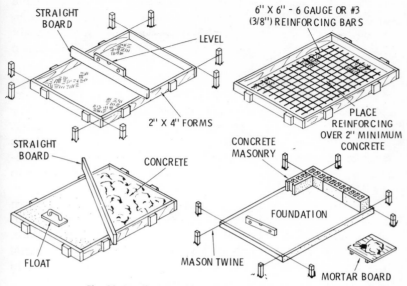

Fig. 13. Details for laying a slab for a barbecue pit.

In areas of severe weather, excavate the earth to a depth of about 7″. Place about 4″ of sand or other granular fill in the excavation—water and tamp thoroughly. Place 2″ × 4″ form boards in the proper location. These should protrude about 1″ above the ground. Set reinforcing bars or mesh in place, about 2″ above the base. Use clean rocks or stirrups to hold the reinforcement. Be sure the forms are level by using a spirit level to check them.

Mix the concrete according to Table 1, or use prepared mixes such as *Sakrete*, for the entire quantity at one time. Pour the concrete between the forms and screed (level) with a straight board. When the water has evaporated, float the surface of the slab and cover with plastic sheeting. Allow the slab to cure for a few days before proceeding with construction of concrete blocks for the pit.

Concrete blocks, as described in Chapter 11, are used to build the barbecue pit. Be sure all necessary material is on hand before starting to build the pit. Standard 3-core stretchers are used, plus end pieces and finished top pieces. Firebrick is used to line the inside of the pit. Extras needed at once are the vertical reinforcing

265

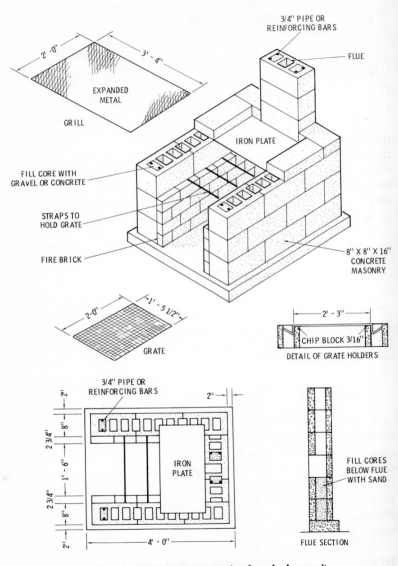

Fig. 14. Concrete block construction for a barbecue pit.

bars or pipe, straps to hold the grate, and the iron plate shown in the sketch. The straps are set in the mortar when the blocks are laid up. Purchase mortar in prepared packages, requiring only the addition of water.

Standard firebrick is 9" × 4½" × 2½". They are laid in place with the wide face exposed to the flame. It takes about 4 bricks to the sq. ft., and their weight is about 35 lbs. to the 100 bricks. For best results, use special air-setting, high-temperature cement mortar.

Chisel out one face to the center core of the third block up, in the back. This will be the entry for smoke into the center core of the 3 blocks acting as a chimney. The iron plate acts as a deflector, directing the fumes up the flue. Apply mortar and lay up the blocks as instructed in Chapter 11. The mortar joints should be the standard ⅜" thickness, which will allow the grate holders and iron plate to be included between the mortar joints. The grate and grill are laid loose across the grate straps and top layer of blocks. This permits removal for cleaning or for storage in winter.

CONCRETE FLAGSTONE WALKS

Natural flat rock is available for placing in the soil for sidewalks. These must be cut to size and shape. The method described here, however, uses concrete. There are two general methods for laying concrete flagstone walks. They are:

1. Placing concrete directly in specially prepared excavations.
2. Precasting concrete in forms of various sizes.

Although both methods have been employed with equal success, considerable trouble may be avoided and the use of forms eliminated by casting the stepping stones directly in their place of usage, especially if they are to be irregular in shape.

If this method of providing flagstones is selected, all that is necessary is to remove the soil by excavating 2" to 4" deeper than the thickness of the slab, so that the walk will have the desired shape and with the edges vertical.

Prior to casting the concrete, a sand base of screened sand is tamped into place to provide a good foundation for the concrete block as well as to furnish drainage. The concrete is then placed and troweled. Colored surface finishes may be employed if desired.

Precasting Flagstone Slabs

No trouble should be experienced in building forms for precast slabs, which consist of a rectangular wooden frame divided by crosspieces or strips of the desired dimensions. Simple forms for casting concrete flagstone slabs are shown in **Fig. 15.**

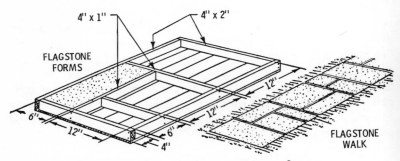

Fig. 15. Form construction for precast concrete flagstone.

As seen from the illustration, different patterns can be made by varying the arrangement of the crosspieces or strips which divide the blocks. The division of the irregular shaped blocks can be made of plastic clay. The blocks should not be made too large to handle since a block 12″ × 18″ × 4″ will weigh approximately 75 lbs.

Before the slabs are laid, a trench 2″ to 4″ deeper than the thickness of the slabs should be excavated. Sand is then screened and rammed in the trench to a depth of 4″ below the finished surface so as to provide for drainage and a smooth base for bedding the stones. The joints are then filled with sand or loam. Where the earth is porous and rather free of large gravel, a smooth bed may be made without sand. The curing of concrete blocks used in flagstone construction is similar to that previously described.

CONCRETE STEPS AND STAIRWAYS

Concrete steps may be divided into two general classifications or types, with respect to their construction, as:

1. Earth-supported.
2. Self-supported.

The earth-supported type shown in Fig. 16 is perhaps the simplest to build since it is entirely supported by tamped earth or gravel. The self-supported type of step, on the other hand, requires reinforcement as illustrated in Fig. 17.

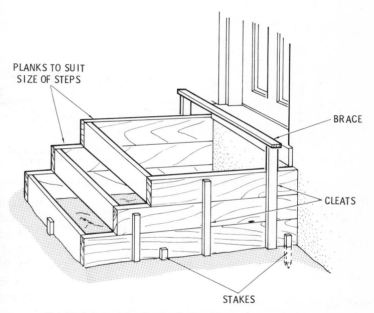

PLANKS TO SUIT
SIZE OF STEPS

BRACE

CLEATS

STAKES

Fig. 16. An example of a simple form for earth-supported steps.

Size of Steps

There are two terms generally used in step construction. They are *riser* and *tread*. By definition, the vertical height or face of a step is called the *riser*, and the horizontal surface the *tread*.

The riser should bear a certain relation to the width of the tread and at the same time should be between 6″ and 8″. Low risers and broad treads are generally preferred for outside steps. Risers from 6½″ to 7½″ high, with properly proportioned treads, permits

269

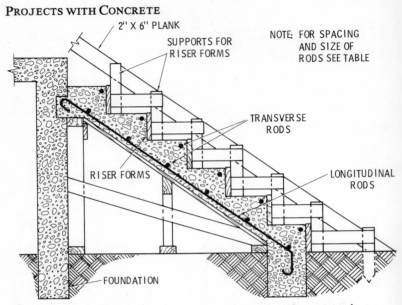

Fig. 17. Steps that are self-supported must use steel reinforcement rods.

walking up or down with the least amount of effort and are most commonly used.

High risers and narrow treads are used only where horizontal distance is limited. A desirable formula to use is that twice the height of the riser plus the width of the tread should equal 25. The treads should have a pitch of from 1/32" to 1/16" toward the front in order to shed water.

Molding Effect

Where desirable, a molding effect may be obtained in the riser by building a form such as that shown in Fig. 18. Another method is to have the riser boards inclined at an angle of approximately 5°, also shown in Fig. 18. The purpose of both is to add a small portion to the actual tread without increasing the overall dimensions of the step.

Forms for Concrete Steps

Where earth-supported steps are to be built, the first move in setting the forms is to prepare the ground over which the steps are

270

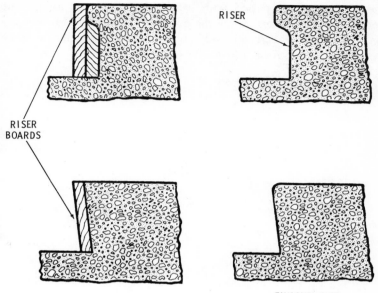

RISER

RISER
BOARDS

FINISHED STEP

Fig. 18. Risers may be molded by a little extra effort in form construction.

to be made. Where the soil is well drained and contains gravel and sand in relatively small amounts, it is only necessary to level and tamp the area thoroughly to compact the soil.

On the other hand, if the soil is a heavy tight clay through which drainage of surface water is slow in wet seasons, it will be necessary to excavate to a depth of 6″ to 8″ below grade and put in a tamped gravel fill. Thorough tamping of the fill is of the utmost importance.

A simple type of concrete step form is shown in Fig. 17. This type of form is made the width of the door frame which it is to serve. In some types of stoops, this width may be necessary or desirable, but the general practice is to set the form to give a width that will extend from 6″ to 8″ on each side of the door frame. The forms for such a stair are shown in Fig. 19. This width will also permit the installation of an ornamental iron railing where desired.

Lumber used for the forms should be smooth on one side and free from knotholes. In addition to adequate staking, the boards

271

should be well supported with braces and stiffeners to prevent bending under pressure when the concrete is poured. This is especially important where the width of the steps makes it necessary to use long riser form boards.

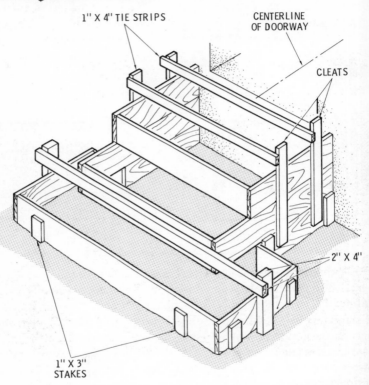

Fig. 19. Concrete steps formed for earth support having an extension base at the foot which could incorporate an iron hand railing.

Figs. 20 and 21 illustrate methods of setting forms for basement stairs. Where the stairwell is already formed with concrete walls, the stair forms are simply riser boards spaced, plumbed, and wedged between the walls. The earth fill is tamped to a uniform slope. When the steps are wider than normal, it is necessary to fit side planks and vertical cleats so that the riser boards may be supported by horizontal braces which are wedged against the side

272

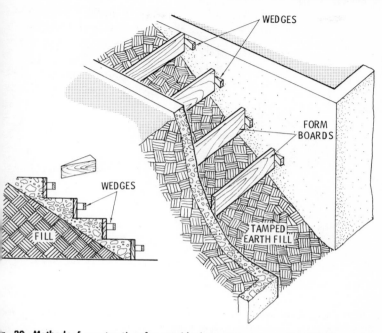

Fig. 20. Method of construction for outside basement stairs wedged between two existing concrete walls.

planks as shown in Fig. 21. Care should be observed in laying out and setting the forms for basement stairs to keep the tread-to-riser proportions within the accepted comfort range.

A typical form of concrete stairs of the self-supporting type is shown in Fig. 18. Steps of this type which do not rest on solid earth or fill must be self-supporting and hence must be reinforced. The longitudinal reinforcement given in Table 3 should be placed lengthwise from top to bottom, 1″ from the underside of the slab.

It is advisable to place rods of small diameter, 12″ to 24″ apart, extending across the width of the slab. These rods should be securely wired to the larger rods at the intersections so as to maintain their position and shape while the concrete is being poured. The forms should be assembled in such a way so as to be sufficiently tight to prevent leakage, and also to facilitate their removal without damage to the concrete. It is very important that self-supporting

273

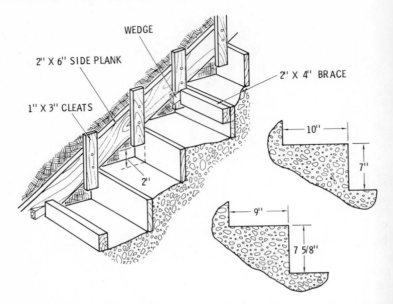

Fig. 21. In the case of wide basement steps, additional side boards and cleats are needed to support the riser boards.

steps have a firm support at the head, such as the slab of a concrete porch or masonry wall.

Table 3. Recommended Dimensions and Reinforcement Rods for Self-Supporting Stairs

Slab dimensions		Round reinforcing rods			
		Longitudinal		Transverse	
Length (ft.)	Thickness (in.)	Diameter (in.)	Spacing (in.)	Diameter (in.)	Spacing (in.)
2 to 3	4	1/4	10	1/4	12-18
3 to 4	4	1/4	5½	1/4	12-18
4 to 5	5	1/4	4½	1/4	18-24
5 to 6	5	3/8	7	1/4	18-24
6 to 7	6	3/8	6	1/4	18-24
7 to 8	6	3/8	4	1/4	18-24
8 to 9	7	1/2	7	1/4	18-24

*ASTM Specifiication C91, Type 1

CHAPTER 15

Tile

Tile is not unlike brick, in its material origins and general method of production. The ingredients are essentially earth clays that are baked hard, like brick. In particular, however, tile is pressed into

Courtesy Tile Contractors Association of America, Inc.

Fig. 1. Reconstructed Egyptian hall showing tile floor.

much thinner bisques and smaller surface sizes and then baked. The baking and face glazing produce a variety of finishes that make tile suitable for decorative purposes. Unlike brick which is used for structural purposes, tile is used to apply a surface to either walls or floors. It has a high glossy surface that is impervious to moisture and the adherence of dirt. It is easy to keep clean.

The dictionary defines tile as being used for floors, walls, roof coverings and drain pipe. Some roof coverings are made of baked clays, the same as the drain tile used underground. This chapter is devoted to the tile used for wall and floor coverings, the installation of which is the work of a mason.

The term ceramic tile is used to distinguish the earth clay tiles from the newer metal and plastic tiles. The use of ceramic tile dates back four to six thousand years ago. It was used in the early Egyptian, Roman, and Greek cultures and, because of its durabil-

Courtesy Tile Contractors Association of America, Inc

Fig. 2. Most common use of ceramic tile is in bathrooms, since tile is not affected by moisture.

Courtesy Tile Contractors Association of America, Inc.

Fig. 3. Tile walls are ideal for atriums where plant moisture is high.

Courtesy Tile Contractors Association of America, Inc.

Fig. 4. A breakfast nook adjacent to a kitchen will last many years, in spite of high humidity.

ity, is sometimes the only parts of structures left to study from the diggings into the ruins of those ancient days. (Fig. 1).

The introduction of ceramic tile in the United States began about the time of the Philadelphia Centennial Exposition in 1876. Up to this time, ceramic tile was made and used in Europe. Some English manufacturers showed tile products at their exhibits, at the exposition, which intrigued some American building material producers. Subsequently a factory was started in Ohio then in Indiana. Since then, of course, its use has grown and there are processors all over the country.

Courtesy Tile Contractors Association of America, Inc.

Fig. 5. A fully tiled public swimming pool.

TILE APPLICATIONS

The uses of tile range from the obvious to the exotic. The obvious begins with its use in a bathroom (Fig. 2), because water spray does not affect it and glazed tile matches tubs and other fixtures of the bathroom. Other areas of the home where moisture may affect conventional wall finishes are likely choices for tile. In the illustra-

278

Courtesy Tile Contractors Association of America, Inc.

Fig. 6. Many deluxe apartments feature all-tile swimming pools.

Courtesy Tile Contractors Association of America, Inc.

Fig. 7. Most ocean cruisers have one or more tiled swimming pools.

279

tion of Fig. 3, a varicolored tile wall is used in an area near heavy indoor plant growth. Kitchen-sink top and back-splashes are a natural for tile. Fig. 4 shows an entire wall in tile, in a breakfast nook. For the same reason, tile is ideally suited to swimming pools. Walls and floors, as well as the pool itself, are covered with ceramic tile in the public pool, illustrated in Fig. 5. Fig. 6 shows a fully tiled odd-shaped pool in a deluxe apartment building. A swimming pool on a deluxe ocean cruiser is illustrated in Fig. 7.

Courtesy Tile Contractors Association of America, Inc.

Fig. 8. A building foyer with tiled floor and walls. Wall design enhances the effect of height.

For reduced maintenance and beautification, tile is an excellent choice for the walls and floors of the entrance foyers of public buildings. In the illustration of Fig. 8, the floor tile is treated to prevent slipping, and is laid with a decorative pattern. The design and pattern of the walls show tile used to enhance the height of the entrance foyer.

Because tile is kiln fired, it accepts heat without effect. Fig. 9 shows a hooded-type indoor fireplace built with brick which is

280

Courtesy Tile Contractors Association of America, Inc.
Fig. 9. A hooded fireplace with wall and base surface of tile.

Courtesy Tile Contractors Association of America, Inc.
Fig. 10. Outside wall of decorative mosaic tile is long lasting and very easy to keep clean.

281

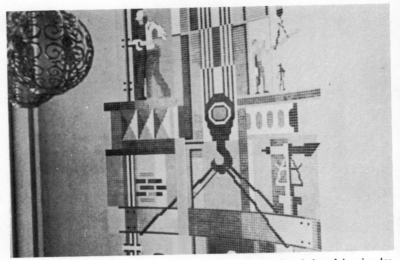

Courtesy Tile Contractors Association of America, Inc.

Fig. 11. Interior hall wall of a trade union building. Tile sizes are the same. Artistry is achieved by varying the colors.

Courtesy Tile Contractors Association of America, Inc.

Fig. 12. Exterior wall tile artistry.

Courtesy Tile Contractors Association of America, Inc.

Fig. 13. Closeup of work shown in Fig. 12. Individual tiles have been cut to proper shape.

Fig. 14. Patterned mosaic tiled walls of a public school. The floor is terrazzo.

faced with tile. Here is a real modern touch to fireplace design, coupled with structural stability. Decorative tile for external wall design may be seen almost everywhere. Fig. 10 is an example. Tile will stay clean and bright year after year.

Perhaps the extreme in decorative and creative tile artistry is shown in Figs. 11 and 12. In Fig. 11, all tile sizes are the same— only the colors vary. The strong artistry and the tile add strength to the story of the building trades. Fig. 12 shows an outside wall, with closeup details in Fig. 13. Here the tiles are precut to shape to form the pattern of this remarkable piece of art work.

The installation of tile is initially a costly operation, mostly in labor. In the long run, however, a tile'wall and floor represents a savings, because of its long life and the low cost of maintenance. Because of this and the versatility of artistic design, architects will specify tile in industrial buildings.

Figs 14 through 17 illustrate the tile treatment of walls in a public school. Note that identical wall patterns are used but with

Fig. 15. The same pattern is maintained throughout the school hallways.

a change in pattern design for the stairway in Figs. 16 and 17. This is the St. Louis Park High School, in Minneapolis, Minn. Fig. 18 shows both the wall and floor of an office building covered with tile. It will stay bright and shiny for years, with simple washing.

Fig. 19 shows a building lobby with a tiled floor that has the appearance of carpeting. This floor will stay new-looking longer, with simple mopping. The most frequent use of tile walls and floors is for public washrooms, in addition to residential bathrooms. Figs. 20 and 21 show two of the public type washrooms. Public washrooms require frequent cleaning and the best way to reduce the amount of work involved is to use tile.

TILE CHARACTERISTICS

Ceramic tile has several advantages for wall and floor coverings, most important of which is durability. They retain their attractive appearance through the life of the structure. A building owner

Fig. 16. A change of tile design is used for the school stairwell.

Fig. 17. Stairwell design is applied to all stair walls. What appears to be a difference in appearance from Fig. 16 is the effect of light reflection at different angles on the tile.

may change a tile wall or floor for a change in design but seldom because the tile has worn out.

Permanency and integrity of color are other advantages of ceramic tile. The bright colors of glazed tile never fade or become muddy—the colors are high-temperature fired into the surface. Dirt, grease, and water spots may accumulate on the surface of tile, but are easily wiped off.

A hard baked clay tile floor covering has great strength and durability. It is easy to install and can be put into almost immediate use after installation. In case of floor damage, individual tiles are easily lifted out and replaced by new tiles. Because of the color permanency of tile, there is generally no problem of matching colors when new tiles replace damaged ones.

The large variety of colors and patterns in ceramic tile makes an endless variety of designs possible, limited only by the artistry

Fig. 18. Tiled wall and floor of a small office building. The cost of the tile installation will more than offset the lower maintenance cost.

Fig. 19. A tiled lobby floor with a pattern that simulates carpeting.

of the architect or tile mason. Increased demands have resulted in some prefabricated panels faced with tile laminated to the structural panels. The laminations include insulation sandwiched between the tile facing and metal, wood, or cement asbestos sheets. Insulating

287

Fig. 20. Nearly all public washrooms have accepted tiling as the standard wall and floor treatment.

Fig. 21. School washrooms are tiled. Wall markings are easily removed.

materials include glass fiber, foamed resins, or foamed glass. In addition, tile is also available already placed on large paper or cloth sheets for easy installation onto walls or floors. Fig. 22 shows a sheet of ¾ " square tile, used for floors and walls. The wall type (4¼ " × 4¼ ") tile is also available in sheets (Fig. 23) as well as individually. Assembled sheets of tile reduces hand labor in placing them. Thus small tiles are applied to a wall in large sheets, from which front supporting paper is removed after the tile is placed into position, or rear supporting material is left on.

Courtesy Tile Contractors Association of America, Inc.

Fig. 22. Small tile pieces, generally known as mosaic tile, are supplied on sheets for easier installation.

CERAMIC TILE USES

Tile is used for two basic purposes, to cover walls and to cover floors. The two types of tile differ and are seldom interchangeable. Wall tile generally has a glazed surface which has been added to the basic tile bisque. Floor tile, for the most part, contains the same material all through its structure and does not include the thin glazed surface. While wall tile is sometimes used for floors where traffic is light, heavy traffic can crack the outer glaze.

TILE

Wall tile are used for interior purposes only and for vertical surfaces, ceilings, countertops and other horizontal surfaces where there are no heavy loads. Wall tile is not weatherproof and is used on exterior surfaces only in frost-free climates. The glaze can be scored and cracked when subject to heavy impact.

Glazed tiles may be used for flooring where traffic is light such as dressing rooms, or the hearth in front of fireplaces. Glazed ceramic mosaic tile is excellent for swimming pools, because of the ease with which they can be cleaned.

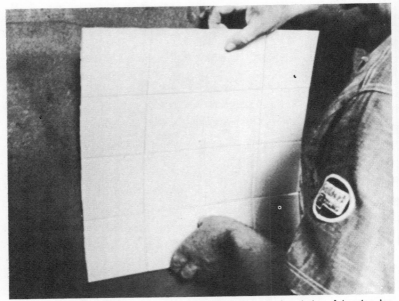

Courtesy Tile Contractors Association of America, Inc.

Fig. 23. Common 4¼" x 4¼" tile may be purchased individually or on sheets.

For floor use, the unglazed tile is preferred. In this tile, the material is the same throughout. These come in a variety of types and colors, but generally in the brown tones, except for porcelains. They may be small mosaic tile, or larger "quarry" (this is the name of a type, not cut from a quarry) tile. They may be had in nonslip types for wet areas, or large sizes for lower labor costs.

TYPES OF TILE

Tile is similar to brick in one respect. It is made of clay, pressed into shape, and fired in an oven like brick. However, since they are not structural, but used for surfacing, the clays are pressed into thin pieces, called bisques. Their thickness varies from ⅜ ″ to ¾ ″, depending on type and purpose. The ingredients vary—from the fineness of the clay dust to the colors of the clays and the color additives. The degree of firing also varies.

In addition, wall tile is made of metal (either aluminum or steel) with a baked on enamel finish. There is also another type of wall tile which is made of plastic. This chapter deals with the ceramic tile because of its close relationship with the products handled by the mason. Basically, there are three types of ceramic tile. The three types are, nonvitreous glazed ceramic wall tile, vitreous ceramic mosaic floor tile, and quarry or packinghouse floor tile used for carrying heavy traffic and heavy loads.

Nonvitreous Glazed Ceramic Wall Tile

These tiles are dry pressed from a soft clay into bisques and then baked. Their base color is a pale yellow or some variation of this color depending on the clay from which it is made. The surface colors are added by a thin glazed surface coating which is again kiln baked to a hard glaze. They are less expensive than the floor tiles described later, but they are not load-bearing and are generally more water absorbing. The most popular size is 4¼ ″ × 4¼ ″, but they are made as small as 3″ × 3″ and up to 8½ ″ × 4½ ″. A variation of this tile is made as large as 8″ × 24″ but their dimensional stability is not too good.

Wall tile is available in a wide variety of colors and patterns. Patterns include geometric patterns, inlays, multiple tones, and even a crystalline wrinkle with a smooth glazed finish. Each color requires separate color addition to the surface and separate heat treatment.

The greatest advantage of the glazed ceramic wall tile is the ease with which it can be kept clean. The colors do not fade and dirt and grime do not stick permanently. A simple washing cleans them like new.

Vitreous Ceramic Mosaic Floor Tile

This tile is made in two general types, a clay type and a porcelain type. Neither have a fired-on overlay of glaze but colors are added to the basic material and appear throughout the structure. As a result, any wearing of the surface leaves no change in the color.

The clay type is made of a finer and denser clay than wall tile, and are fired at a higher temperature. As a result they are less water absorbent (only about 1%) and may be used for exterior purposes. Earth dyes are included in the clay at the time of pressing. New manufacturing processes permit mottling and veining effects of the color as well as the solid colors.

The porcelain tile is a denser formulation and the results are a purer color and slightly greater durability. Porcelain tile is made in sizes up to $2'' \times 2''$. These are sometimes called pavers. They were more popular some years back in hexagonal shapes, and were used for bathroom floors.

The term mosaic used here must not be confused with the mosaic murals done by artists for wall decoration. The mosaic-mural artists use what is rightfully called smalti, made of small pieces of colored glass or stones, which is not the same as the mosaic tile described here.

Quarry and Packinghouse Tile

These tiles are made in larger and thicker pieces, than the floor tile, and are made denser and stronger. They are used for carrying heavy traffic and heavy loads. Quarry tile is about $\frac{1}{2}''$ thick, made from reddish brown earth clays and with some color added. Packinghouse tile is about $1\frac{3}{8}''$ thick and carries the color of the natural clay only—no color is added.

In both types, water absorbency is very low, and resistance to abrasion is high. The tough, smooth surface will not carry odors from contamination, grease, dirt, acids, or moisture. These tiles are finding wide favor in home terrace paving. Their natural earth colors give them a "Western" look which fits into the decor of ranch style and modern home styling.

Floor tile specifications depend on the amount of traffic and the load the tile must bear. The thicker the tile, the better is its ability

to handle heavy traffic and heavy loads. A balance must be made
between the greater variety of colors available in the thinner tile
and the limited earth colors of the thicker tile.

Mosaic tile is generally only about ¼″ thick. It offers the architect a greater variety of color and design effects but cannot take the heavier traffic and loads of thicker tile. The pavers are about ½″ thick and can be obtained in about the same variety of colors and designs as mosaic tile. Quarry tile comes in a limited range of

Courtesy Tile Contractors Association of America, Inc.
Fig. 24. Full-mortar-bed on top compared with thickness of thin-bed on bottom.

293

colors but can take more load. At the other extreme, packinghouse tile can take the greatest punishment but is available only in the earth red of its clay.

GENERAL INSTALLATION CONSIDERATIONS

Two basic bedding or backing systems are used for setting tile. The conventional method is called full-mortar-bed. This consists of a bed of standard portland cement and sand mortar laid wet as with concrete or brick mortar. The newer and faster method is called thin-bed which has equal strength and is dry-cured. Both are shown in Fig. 24. On the left is the wet full-mortar-bed and on the right is the thin-bed or dry-mortar bed. This thin-bed has the bonding strength of the thicker conventional mortar bed.

In either type of base, a thin coating is required over the surface for an adhesive for the tile. For the full-mortar-bed system, a thin neat coat of mortar is generally used. For the thin-bed system,

Courtesy Tile Contractors Association of America, Inc.

Fig. 25. There are many types of adhesives (mastics) and grouts, each intended to meet a specific purpose

there are a number of adhesives which include organic ingredients that make them impervious to chemical action—a claim which cannot be made for conventional mortar.

When tile is set in place it must be grouted. This means the spaces between the tile are filled with a bonding agent. Conventional grout is cementitious. Although slightly more expensive, newer grouting material prevents damage by chemicals. The market includes a number of bedding, adhesive, and grouting materials (Fig. 25). The mason or contractor should check various products to conform to the ASASI standards, to assure a quality product.

Courtesy Tile Contractors Association of America, Inc.

Fig. 26. Full-mortar-bed thickness varies with tile size and traffic to be carried.

FULL-MORTAR-BED METHOD

The full-mortar-bed method is especially adapted to floors (but is also used for walls) which must carry heavy traffic. Floors must include a firm foundation to carry the traffic, plus a ¾" to 1¼" setting bed that must be carefully leveled. Fig. 26 shows a cutaway view of the bed for two types of tile. On the right is the larger packinghouse type of tile, with a heavier bed. On the left is 4¼ " × 4¼ " tile, with a thinner bed. The latter is for light traffic, such as residence patios. Fig. 27 is a closeup view of the square tile and its bed. A properly installed bed for floors should also include reinforcing wire mesh, to be described later.

The conventional wet-mortar bed may be used on walls. Fig. 28 illustrates a sketch of a cross-section view. This installation is of the interior wall on wood studs. The treatment is something like stucco. The basic bed is not over ¾ " thick, followed by a plumb coat, then a 1/32" to 1/6" bonding coat and a neat coat of cement or a special adhesive.

Courtesy Tile Contractors Association of America, Inc.
Fig. 27. Closeup of 4¼" x 4¼" tile shown on left in Fig. 26.

THIN-BED METHOD

The thin-bed method is preferred for wall tile installations. It is faster to apply and is as strong as the full-bed methods. It may also be used for floors in light duty applications (Fig. 29). The most versatile thin-bed material is a *Dryset* mortar which is a mixture of presanded portland cement and organic additives which enhance the water-retention characteristics and improve workabil-

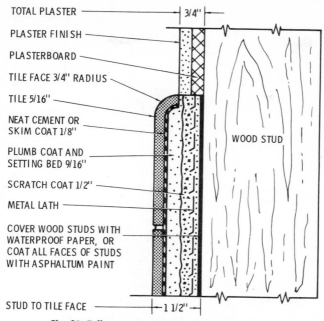

TOTAL PLASTER —————— |← 3/4" →|

PLASTER FINISH —————

PLASTERBOARD —————

TILE FACE 3/4" RADIUS ——

TILE 5/16" ———

NEAT CEMENT OR
SKIM COAT 1/8" ————

PLUMB COAT AND
SETTING BED 9/16" ———

SCRATCH COAT 1/2" ———

METAL LATH ————

COVER WOOD STUDS WITH
WATERPROOF PAPER, OR
COAT ALL FACES OF STUDS
WITH ASPHALTUM PAINT

WOOD STUD

STUD TO TILE FACE —————— |← 1 1/2" →|

Fig. 28. Full-mortar-bed on wood studs for wall tile.

Fig. 29. Thin-bed used in light-duty floor tile installation.

ty. It has the same appearance as full-bed mortars but it is applied
:asier and faster.

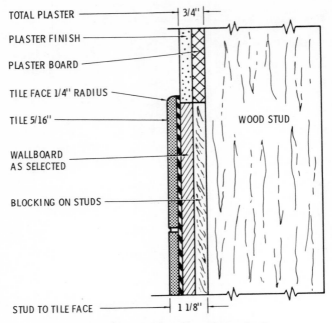

Fig. 30. Thin-bed tile wall installation details.

Another thin-bed type of material uses solvent-based organic adhesives which are quick-curing and can be applied over flat smooth surfaces such as wallboard, metal, and wood paneling. Still another type is a two-part system consisting of an epoxy resin and a catalytic hardener. It is highly resistant to alkalies and acids. It must be applied with a rubber-faced trowel. The illustration in Fig. 30 shows a thin-bed system on wallboard. The stud-blocking layer may be omitted.

GROUTS

Grouting is filling in the spaces between tiles much like the mortar used between bricks. The most commonly used grouting material is cementitious, made of portland cement and very fine white sand. This is very acceptable grout for areas where there is not an excessive amount of moisture and where there is no ex-

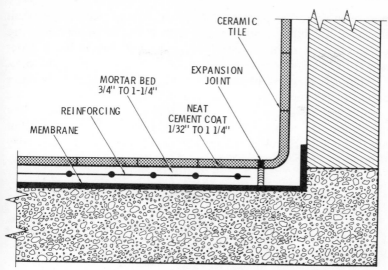

CERAMIC
TILE

EXPANSION
JOINT

MORTAR BED
3/4" TO 1-1/4"

REINFORCING

NEAT
CEMENT COAT
1/32" TO 1 1/4"

MEMBRANE

Fig. 31. Construction details of a full-bed tile floor installation.

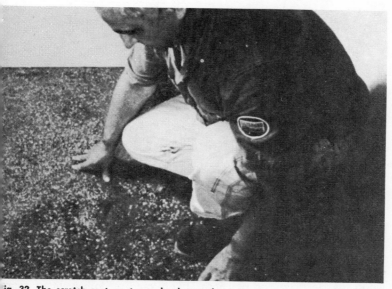

ig. 32. The scratch coat must cure hard enough to carry the mason's weight before
proceeding.

299

posure to corrosive chemicals. Food processing plants, restaurant kitchens and other places where animal or vegetable acids are present can result in failure of the grouting if mortar based grout is used. This may be true, as well, in home kitchens and bathrooms.

Although the cost of material is slightly higher, the corrosive action of chemicals can be overcome by the use of one of the organic grouts. These are the epoxy and furan resin types. These grouts are structurally as effective as the mortar-based types, but are immune to chemical action. The grouts mentioned above may be used on floors or walls.

Fig. 34. Separate pieces of mesh must overlap.

CERAMIC FLOOR TILE

Whether for floors or walls, meticulous preparation for smoothness is required. The final job must have tiles securely in place, evenly spaced, and absolutely flat. The preparation of the floor bed will depend on whether the tile floor substrate is gravel, dirt, wood, old terrazzo, or tile floor. If on a gravel substrate, the requirements include a solid concrete base usually followed by the full-bed wet mortar bed for the tile. If on an already solid flooring, either the full-bed or the thin-bed method, which is faster and easier, may be used.

Fig. 31 shows the essential layers of a full-bed floor over a gravel substrate. It begins with a waterproof membrane of plastic sheeting or tarpaper. This is to prevent absorption of the water from the mortar bed, to allow proper curing. The mortar scratch coat, about ¾ " to 1¼ " thick, is poured and allowed to harden slightly, then scratched, for roughness, with a notched trowel. The thickness mentioned is for residential use, with light traffic loads. For industrial floors, it should obviously be much thicker and poured with a low-water-formula concrete. Be sure the basic coat is carefully levelled.

When the scratch coat is hard enough to bear the mason's weight (Fig. 32), a wire mesh is laid down. This reinforces the bed (Fig. 33). The reinforcing mesh comes on long rolls and must be cut to lengths. Edges must overlap (Fig. 34) to assure a continuous

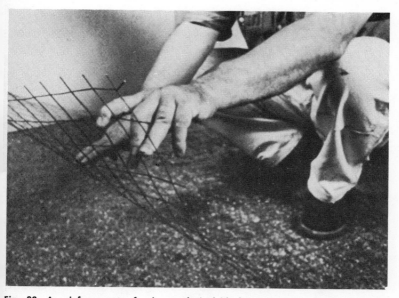

Fig. 33. A reinforcement of wire mesh is laid down to go between layers of mortar.

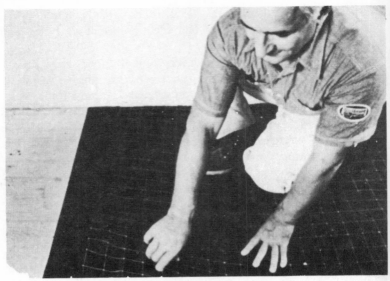

Fig. 35. Bend mesh to lie flat on scratch coat.

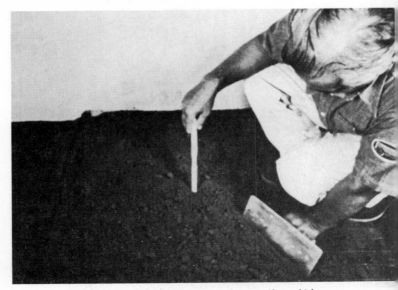

Fig. 36. The final bed coat is poured to a uniform thickness.

Fig. 37. Draw a straight board across the bed coat to level it.

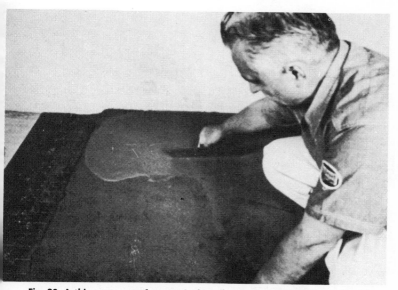

Fig. 38. A thin neat coat of mortar is the adhesive for holding tile in place.

mesh reinforcement. Be sure the mesh is laid out straight (Fig. 35), since none of the wires must protrude through the covering mortar coat, which is comparatively thin.

Dampen, but do not soak, the scratch coat with water. There must be no standing pools of water. Pour another layer of mortar over the first, about ⅜ " thick (Fig. 36). Screed the top (level it) with a straight board. This is probably the most important step for final tile flatness (Fig. 37). Before the bed of mortar has fully cured, the mortar for the adhesion of the tile is troweled on. This is called the neat coat (Fig. 38) and is only 1/32" to 1/16" thick. Tile is then placed over the neat coat, pounded into place and grouted. This will be discussed later.

Instead of the neat coat as an adhesive, the mason may use a thin layer of one of the many dry-curing mortars, such as *Dryset*. This has better bonding action than wet mortar and the installation is faster, allowing time for more careful spacing of the individual tiles. However, the bed on which it is applied must be perfectly flat. Tile is not easily pounded flat after laying.

Fig. 39. A notched rubber trowel is used to spread one of the organic mastics.

A number of organic adhesives are available for bonding the tile. Some are made especially for nonvitreous tile and others for vitreous tile. These adhesives are not recommended where excessive dampness is encountered. The best tile adhesive, even though more expensive, is one of the epoxies. They have the best bonding qualities and may be applied in wet areas without affecting their quality. They are spread with a notched rubber trowel (Fig. 39). The organic adhesives and sometimes the cementitious types are often referred to as *mastic*. This is the term usually used for do-it-yourself homeowners.

Fig. 40. Nonvitreous tile must be soaked in water before use.

LAYING IN THE TILE

Tile is supplied in two forms. The larger sizes which are the nonvitreous wall tile with a glazed face, and the vitreous quarry and packinghouse tile which comes in individual pieces. Smaller tile (the vitreous mosaic) is always supplied on sheets for easier handling. Individual mosaic tile would take many hours to install otherwise.

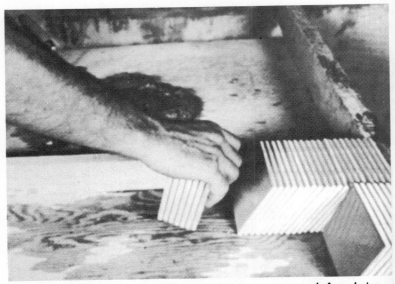

Fig. 41. Tile, stacked near the work area, is draining excess water before placing.

Fig. 42. Individual tiles are placed carefully onto the adhesive or mastic.

Nonvitreous tile will absorb water. If laid on a cementitious neat coat, they must first be soaked in water (Fig. 40) to prevent drawing moisture from the mortar. The soaking should be started several hours before installing the tile. When ready to use, stack them near the work area (Fig. 41) to permit drainage of excess water. Pick up a few at a time and place them carefully as shown in Fig. 42. The holding paper of tile in sheets is conventionally on the front, or top side, of the tile (Fig. 43). These are vitreous tile and need no soaking.

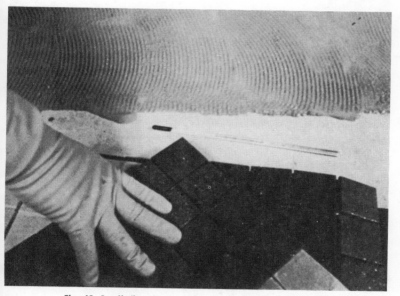

Fig. 43. Small tile, glued to paper sheets, are easy to install.

It is usually advisable to smear spacer mix over the back (bottom) and work it into the spaces between the tiles. Spacer mix is similar to grout. Work the mix in, but be sure all of the mix is off of the back (bottom) surface. Place the sheets of tile over the adhesive (Fig. 44), being sure to place sheets next to each other to maintain the same spacing and alignment.

A recent trend among suppliers of glazed wall and mosaic tile has been to supply sheets of tile with the bonding fabric on the

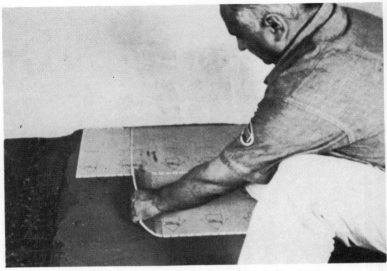

Fig. 44. Careful alignment of tile in sheets is important.

Fig. 45. Beating the tile level with a wood block and hammer.

back (or bottom) instead of the front. This eliminates the need to remove the bonding material. The bonding material may be perforated paper, or loosely woven plastic mesh. Due to the open construction of the material, the plastic adhesive or mastic easily reaches the backs of the tile for a secure hold.

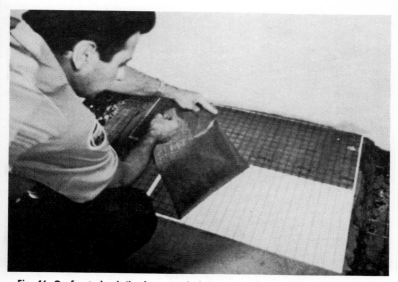

Fig. 46. On front-glued tile sheets, soak the paper with water, then remove paper.

PLUMBING THE TILE

Plumbing the tile is an important operation in getting a smooth and flat surface to the finished job. Whether the tile is placed in sheets or individually, use a piece of wood and a hammer to smooth the surface. Place the wood block over the tile, and tap the back of it with the hammer (Fig. 45). Move the block around over the entire surface, tapping away with the hammer as you go from place to place. When done, you should be able to run your hand across the tile face and feel a perfectly smooth surface with no bumps or depressions.

Organic adhesive or mastic may squeeze through the edges of tiles and smear the front surfaces. A special solvent for these types

309

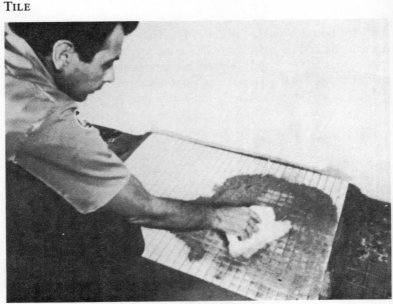

Fig. 47. Wash glue off tile face.

of adhesives is available for cleaning the face of the tile, before the grout is applied.

GROUTING

Grouting can begin after plumbing, except that tile laid from sheets must be allowed to set for about an hour before removing the holding paper, if front glued tile is used. Soak the paper to loosen the bond. Pull the paper off (Fig. 46). Using water, thoroughly wash off the remaining glue from the tile (Fig. 47).

Apply cement type grout with a rectangular trowel (Fig. 48), working it into all of the spaces thoroughly. If one of the organic grouts or epoxy, is used, spread it with a rubber-edged trowel (Fig. 49). Clean off excessive grout of the organic type with rags. Organic grout is also easily applied with a window squeegee. Use the rubber of the squeegee to force grout into the spaces between tiles. Run the squeegee across the face of the tile to wipe off excess, followed by a sponging.

Fig. 48. Spread cementitious grout with the edge of a trowel.

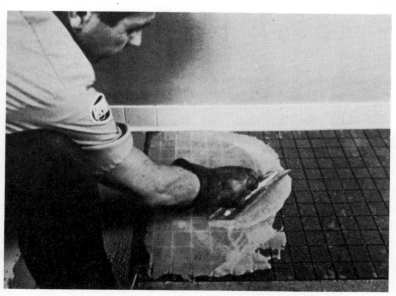

Fig. 49. A rubber-faced trowel is used to spread organic type grout.

For cleaning the cementitious grout, brush off as much of the excess as possible. For glazed tile, allow two days for curing then wash off all excess grout. Unglazed tile must be allowed to cure for 10 days. It is then necessary to remove surface grout with a dilute acid solution. Make a 10% solution of muriatic acid and wash the tile faces. Rinse off the acid and wash with an abundance of water.

WALL TILE INSTALLATION

Wall tile installation is basically not too unlike that for floor tile. Three methods of installation are in common use. The full-mortar-bed method uses a comparatively heavy bed of portland cement mortar and the grout used between the tiles is also a portland cement mix. The method is permanent and sturdy and may be used on any type of wall except very smooth surfaces such as *Masonite,* plaster, plywood, metal, etc.

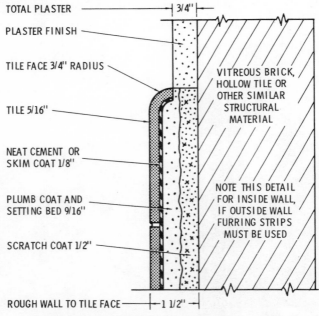

Fig. 50. Full-mortar-bed tile details over a vitreous solid wall.

The second is the thin-bed system, used on lightweight wall construction. It features high-speed installation and resistance to chemicals and other adverse environments. The mortar, one of which is *Dryset,* includes organic additives in a presanded portland cement. The additives enhance water-retention characteristics and make the mortar more workable. It has the same handsome appearance as full-bed mortar but lays in easier and faster. It is used on the same type of surfaces—over concrete, cinder block, brick, tile, etc. The grout used with thin-bed mortars is specially formulated and is a self-curing dry tile grout similar to *Dryset.* It includes additives which make it stain resistant and reduce powdering.

The third method employs a special adhesive for installing tile to smooth surfaces. It is not really a bed but an adhesive only. However, it does adhere to wallboard, plaster, plywood, metal, etc. The surface must be perfectly smooth.

Fig. 28 shows the full mortar bed method used on wood wall studs. Fig. 50 shows the layers over vitreous solid walls. This method also applies to concrete and concrete block, brick, etc. A

Courtesy Tile Contractors Association of America, Inc.

Fig. 51. Applying the first, or scratch, coat on a well.

313

Courtesy Tile Contractors Association of America, Inc.

Fig. 52. Scratch coat over a concrete wall.

Courtesy Tile Contractors Association of America, Inc.

Fig. 53. Scratch coat over a cinder block.

Courtesy Tile Contractors Association of America, Inc.
Fig. 54. Level the scratch coat by drawing a straight board (screed) over the surface.

Courtesy Tile Contractors Association of America, Inc.
Fig. 55. Scratching the surface with a piece of metal mesh.

315

scratch coat is first applied (Fig. 51) in much the same way as applying stucco to the outside of a home. If on wood studs, the tarpaper or other waterproof membrane and the wire mesh is in place first. This coat should be about ½″ thick but should not be over ¾″. On concrete or cinder block surfaces (Figs. 52 and 53) no wire mesh is needed.

The scratch coat is allowed to damp cure for a short period of time after which a screed or leveling board is drawn across (Fig. 54) to make sure it is perfectly level. The coat is then scratched (Fig. 55) to make a rough surface to bond the bed coat. Use a scrap piece of the metal mesh or the toothed edge of a trowel.

The scratch coat is then wet down, but not saturated, with water (Fig. 56) after which the setting bed is applied. The setting bed is the same as the scratch coat, between ½″ an not over ¾″ thick. The setting bed may be used in two ways. It may be allowed to cure dry and a thin-bed applied later or a thin neat coat is applied while the setting bed is still wet. The neat coat should only be 1/32″ to 1/16″ thick, and is the coat to which the tile adheres.

Courtesy Tile Contractors Association of America, Inc.

Fig. 56. Moistening the scratch coat before applying the bed coat.

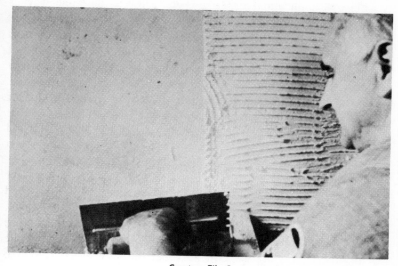

Courtesy Tile Contractors Association of America, Inc.

Fig. 57. The adhesive coat, whether neat coat, organic adhesive, or mastic, is only ⅓₂" to ⅟₁₆" thick.

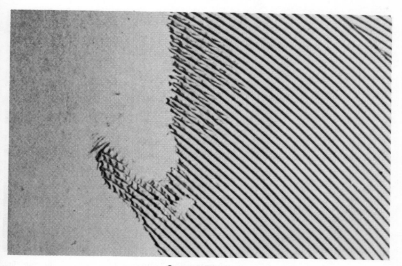

Courtesy Tile Contractors Association of America, Inc.

Fig. 58. Organic adhesive or mastics will adhere to plasterboard.

Thin bed application for walls is the same as described earlier for floors shown in Fig. 30. One of the adhesives mentioned is applied over the thin bed (Fig. 57) and grooved with the notches of the trowel. Tile is placed onto the adhesive. Modern internal wall construction has gone more and more to the thin-bed systems because new products developed have excellent bonding qualities.

There is less expense because less labor is involved and it permits constructing thinner walls which leave more space for room sizes.

ADHESIVES

The conventional adhesive is a thinner mixture of portland cement mortar and is given the name neat coat. A neat coat may also be used over the thin-bed type surfaces. While their ability to hold tile is excellent, they are not immune to the action of chemicals which may penetrate the grout. Organic adhesives have been developed that are immune to alkali and acid action.

Courtesy Tile Contractors Association of America, Inc.

Fig. 59. For tile directly on wood, use organic type adhesives only.

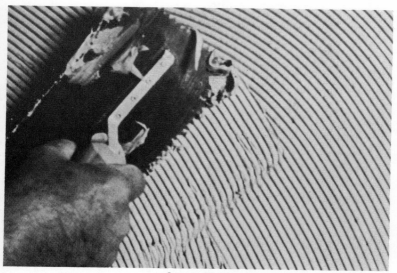

Courtesy Tile Contractors Association of America, Inc.
Fig. 60. All adhesives should be applied with a toothed trowel.

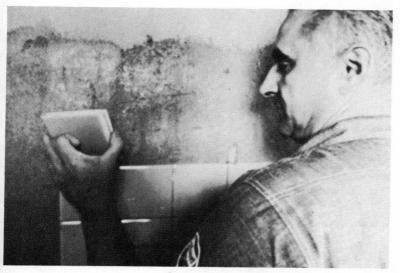

Courtesy Tile Contractors Association of America, Inc.
Fig. 61. Laying tile against the mastic.

Furan resin, latex resin, and the epoxies, are immune to chemical action. The best of these is epoxy but it is higher in cost and is uneconomical to use unless its special qualities are needed. The organic adhesives may be used over smooth surfaces, such as plasterboard (Fig. 58), plywood (Fig. 59), metal, etc. These adhesives are applied with a notched trowel (Fig. 60) the same as with neat cement. The epoxy adhesive requires the use of a rubber trowel.

It is obvious that when an adhesive is used without a bed coat, such as directly over plasterboard, the surface must be smooth and level. There is no opportunity to plumb the wall surface as with a mortar-bed coat. The organic adhesives are not recommended for use in areas that are wet or overly hot, during the time of application.

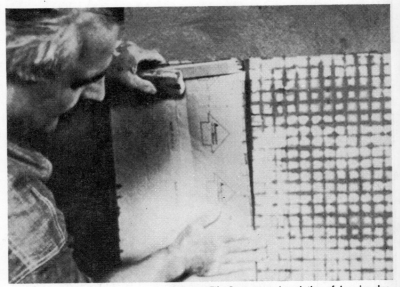

Courtesy Tile Contractors Association of America, Inc.
Fig. 62. Carefully align tile sheets edge to edge.

If the tile is nonvitreous, such as the bisque of glazed tile, it must first be soaked in water as described for floor tile. The tile is available individually or on sheets, depending on the size of

Courtesy Tile Contractors Association of America, Inc.
Fig. 63. Set up a vertical guide line.

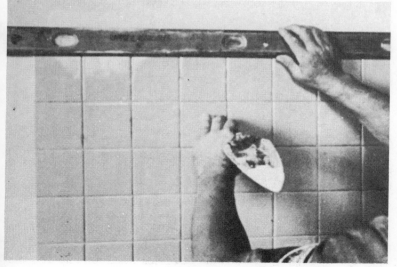

Courtesy Tile Contractors Association of America, Inc.
Fig. 64. Check rows frequently with a carpenter's level.

321

the tiles. When placing them individually (Fig. 61), extra care is necessary to be sure they are evenly spaced. Tile on sheets (Fig. 62) reduces the tediousness of precise laying. Set up vertical guides (Fig. 63) or use the outside or inside corners of the room. As the rows are laid up, use the spirit level frequently as illustrated in Fig. 64.

Since mortar has a tendency to shrink as it cures, expansion joints must be cut into the mortar bed every 17" to 24" (Fig. 65). Use the edge of a trowel and make the cut through the vertical joint of the tile. Fig. 66 shows the cut being made through a sheet of tile.

Sheets of tile should have a bit of damp spacer mix forced between the joints, on the back side. This helps maintain the spaces between tiles after the paper is removed. Press the sheets into place, using care to align each sheet, edge to edge. Before the adhesive has had a chance to cure and become hard, the tile must be beat with a block of wood and a hammer. Move the wood block over the surface of the tile (Fig. 67) as the wood

Courtesy Tile Contractors Association of America, Inc.

Fig. 65. Cut expansion joints through tile joints and into the wet mortar bed.

Courtesy Tile Contractors Association of America, Inc.
Fig. 66. Cut expansion joints through the paper sheet, on sheet tile.

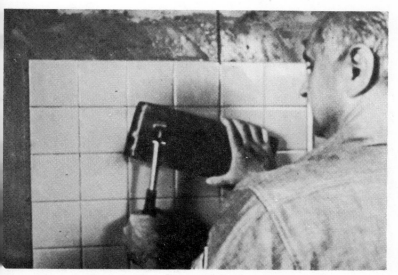

Courtesy Tile Contractors Association of America, Inc.
Fig. 67. Beating individual tile to obtain a level surface.

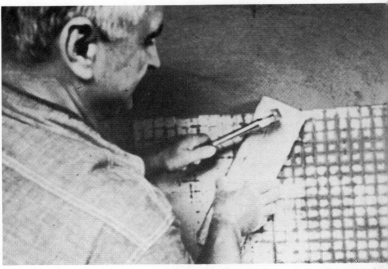

Courtesy Tile Contractors Association of America, Inc

Fig. 68. Sheet-mounted tile, with supporting paper on front, is beat before the paper is removed.

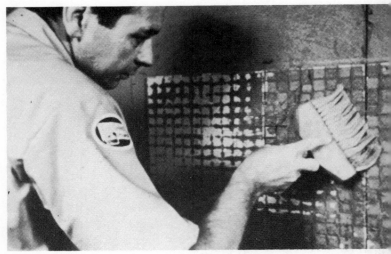

Courtesy Tile Contractors Association of America, Inc

Fig. 69. Wetting down supporting paper on front-mounted sheet tile.

block is tapped with the hammer. The wood block should be large enough to cover several tiles at one time. Tile on sheets is beat before the paper is removed (Fig. 68).

Within an hour after sheet tile has been laid, the backing paper is soaked (Fig. 69) and stripped off (Fig. 70) and the glue completely washed off of the tile (Fig. 71). Grouting begins immediately after the tile is placed. The grout mix is spread over the tile and forced into the joints (Fig. 72). If the tile is square-edged, the grout should fill the joints even with the level of the tile surface. If the tile has curved edges, the joints are tooled (Fig. 73) with a concave tool made for this purpose.

Courtesy Tile Contractors Association of America, Inc.
Fig. 70. Removing soaked paper.

Glazed tile is cleaned with a damp sponge (Fig. 74). The sponge must be damp, not wet. Go over it several times until there is no grout residue left on the surface of the tile. The grout will cure for use in two days. Grout residue will not wash off non-glazed tile. Allow the tile joints to cure for 10 days. Wet the tile face thoroughly, then wash with a 10% solution of muriatic acid

Courtesy Tile Contractors Association of America, Inc.
Fig. 71. Wash off glue with a rag or sponge, and water.

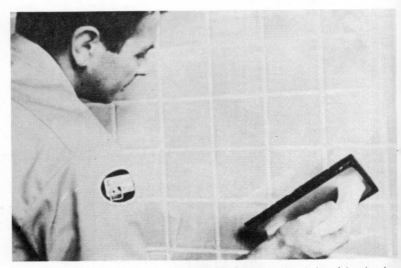

Courtesy Tile Contractors Association of America, Inc.
Fig. 72. Spreading cementitious grout with a metal trowel. Organic grouts are spread with a rubber trowel or squeegee.

Courtesy Tile Contractors Association of America, Inc.
Fig. 73. Tiles with curved edges must have the joints tooled.

Courtesy Tile Contractors Association of America, Inc.
Fig. 74. Wash grout residue off tile with a damp sponge.

which is the only thing that will loosen and wash off cementitious material. When the tile is completely clean, rinse the acid off the tile with water.

Also available from tile suppliers are soap dishes, towel racks, and other accessories to cement into the walls as part of the installation. Sizes are such as to exactly replace one or more tiles. Special finishing tile may be purchased for bottom coving; cap and bullnose tile for the top row; corner tile and narrow pieces for edging.

There are several methods for cutting tile to fit around pipes or other obstructions. While tile is brittle, it may be cut with a hack saw. Contractors use a powered abrasive wheel. A chipper tool, that looks like a large pincher plier is used to chip off small pieces. For breaking a tile in half, use a glass cutter to score across the glazed surface then break the tile, much like cutting glass. Edges of cut tile can be smoothed with medium carborundum paper.

Fig. 75. An aluminum tile kitchen back-splash using standard 4¼" x 4¼" size.

METAL AND PLASTIC TILE

While metal and plastic tile is sometimes used by professional tile installers, their biggest feature is their ease of installation. Their appeal is more to the homeowner who likes to do his own installation. They are slightly more expensive than ceramic tile and, because they are flexible, they can be easily dented.

Metal and plastic tile is intended for installation directly on plasterboard, and is frequently used in kitchens as a back-splash, where a plastered wall is already in place (Fig. 75). No bedding is used, nor are they grouted.

Metal and plastic tile is easily cut for placement around pipes or other obstructions. The illustration in Fig. 76 shows an aluminum tile being cut with tin snips. Plastic tile is cut in the same way. Plastic tile should be warmed to make it pliable, to permit handling without breaking.

TILE TERMS AND DEFINITIONS

Tile—A designation for all types of glazed and unglazed tiles, made exclusively from clay, with or without other ceramic mate-

Fig. 76. Aluminum tile is easily cut with tin snips.

rials, and burned or baked in the course of manufacture. They are made in sizes which can be handled conveniently, installed individually or in multiple units. They are used exclusively as facing or surfacing for interior or exterior walls, floors, pavement, ceilings, etc., and are not a structural unit.

Cement—Cement for use in mortars, setting beds, grouting, and all other operations in connection with the tile work refers to a standard grade portland cement of domestic manufacture. All portland cement shall be of a color other than white unless otherwise designated.

Lime—Lime is used in mortars, setting beds, and all other operations in the tile work. Refers to a high-calcium type lime putty as prepared from domestic hydrated lime or properly slaked quicklime.

Mortar—A mixture of cement (with or without lime putty), fine aggregate, and water, with or without the addition of admixtures such as integral waterproofiing compounds.

Aggregates—The term "aggregates" applies to the noncementitious materials such as sand in tile mortars and sand, gravel, and stone used in concrete fill and other operations in connection with the tile work.

Concrete Fill—The layer or bed of concrete serving as the underfill and base for floor tile and other horizontal tile installations. It is applied to the subfloor with a thickness sufficient to bring the mortar setting bed and tile to the proper surface level.

Cleavage Plane or Cushion—A layer of material, such as sand or building paper, used to separate the concrete fill or mortar setting bed from the subfloor or other structural base, to permit a possible independent movement of the subfloor or structural base.

Mortar-Setting Bed—The layer or bed of mortar applied to the concrete fill or other subfloors and subwalls upon which the tiles are laid and beaten in.

Scratch Coat—A scratch coat is a mortar bed applied directly to the metal lath, hollow tile or other types of subwalls for the purpose of providing a sufficiently strong and rigid surface upon which the mortar setting bed or the plumb coat can be properly applied. It also acts as an absorption regulator when applied to bases of a porous nature. The scratch coat is scored, scratched or grooved, to ensure a key and bond with the mortar setting bed.

Buttering—A method of applying the mortar setting bed to the tile before placing the tile and mortar to the scratch coat. In this method, a sufficient quantity of mortar is spread on the back of each tile unit, so that when the tile and mortar are placed upon the scratch coat, the tile can be tapped to the required surface level.

Floating—A method of applying the mortar setting bed to the scratch coat for placing the tile. In this method, the mortar is troweled or spread on the scratch coat to a true plumb surface and is of sufficient thickness to bring the subsequently applied tile to the required surface line. A thin skim coat of neat cement is spread on the mortar bed or on the back of each tile unit. The tiles are beaten into the mortar bed to the required finished surface line.

Screeds or Screed Strips—A wooden strip section or a part of mortar laid on a floor or wall at intervals to gauge the thickness of setting beds or to indicate the finished tiled surface.

Grouting and Pointing—The method of finishing the joints of all tile work by filling them with cement or a mortar.

Inserts—Isolated or grouped tiles, plain or decorated, set in a finished surface of tile or other materials, such as brick, stone, stucco, concrete or cement, whether on interior or exterior walls or floors.

Trimmers or Trim—Bases, caps, corners, angles, architraves, and other tile moldings shaped to suit various types of tile.

TYPES OF TILE

There are two general kinds of tile—glazed and unglazed. Each of these is composed in various types.

Shapes and Sizes—The shapes and sizes of tile available allow wide architectural freedom and design, but for economy and convenience the industry has adopted certain approximate standards which can be furnished by most manufacturers. Individual manufacturers make additional sizes and shapes which can be purchased as special tile.

DEGREES OF VITRIFICATION OF TILE

Impervious—That degree of density of either the tile glaze or the body which will not permit the absorption of any liquid or grime, and from whose surface any stains or grime may be easily removed.

Vitreous—That degree of density of a tile body which will absorb less than 3% of moisture and which will not permit any grime penetration into the face of the tile which cannot be readily removed.

Semivitreous—That degree of density of a tile body which will absorb more than 3%, but less than 7% of moisture.

Nonvitreous—That degree of density of body which (although having high strength) will absorb more than 7% of moisture.

GLAZED TILE

In general, glazed tile has a glass-like surface composed of ceramic materials fused upon its surface. Such glazed surface is impervious. It does not absorb stains nor change color. Ink, pencil markings, oil, and grease may be easily removed. Ordinary acids will not injure glaze except as stated by the manufacturer.

The body of the tile may be composed of white or colored clays or other ceramic materials. It may vary in degree of vitrification or absorption according to the purpose for which the tile is intended. The body of the tile, its color, its method of manufacture, and the glaze applied thereon may be produced by various methods suitable to the individual manufacturer.

Color and Finish of Glaze—The glaze on tile may be clear, opaque, white, black, colored, or polychrome with smooth, mottled, veined, or ripple effect. The finish of the glaze may be bright, semimatte or matte. Each glaze has its own characteristic finish, which is an inherent part of the glaze and cannot be changed readily to meet personal preference.

Bright, Semimatte and Matte Glazes—The surfaces of glazed tile vary in light reflecting properties. If the glazed surface reflects an image, it is called a bright glaze. A glaze surface which does not reflect an image is called a matte glaze. Surface finishes between a bright and a matte glaze are termed semi-matte.

A method of determining whether a glaze is bright (reflecting) or matte (nonreflecting) is as follows: Place a piece of tile in a vertical position on a horizontal surface with its face parallel to the direction of the strongest source of light. Place a white envelope on the horizontal surface 2″ in front of the glazed face of the tile. If the outline of the edges of the envelope are visible on the face of the tile, it is termed a bright glaze. If the tile does not reflect the image of the edge of the envelope, it is termed a matte glaze.

Proper use of glazes should be recognized. Bright surface glazes may be used on walls, but should not be used on floors at any time. Semimatte glazes may be used satisfactorily on interior wall surfaces, and on floors subject to limited floor traffic. Matte-finish glazes, of the proper body, can be used for interior walls and floors restricted to reasonable residential wear.

Weatherproof—The word weatherproof as used, means that representative samples of tile to which the name applies will withstand the standard freezing tests used by the Tile Industry Research Bureau without distintegration of the tile.

Glazed Interior Tile—Symbol GI. This designates a durable tile having an impervious glazed face with a white or colored body composed of clay or other ceramic materials. This type of tile is suitable for use in interior locations not subject to freezing temperatures in the presence of water. The thickness of this tile shall be at least ¼″, for units up to 36 sq. in. in facial area, and at least ½″ for larger units.

Glazed Weatherproof Tile—Symbol GW. This designates a tile having an impervious glazed face and a semi-vitreous or vitreous white or colored body composed of clay and other ceramic materials, and which will pass the standardized freezing tests as established by the Tile Industry Research Bureau. The thickness of this type of tile is usually ½″, but it may be a minimum of ⅜″, if less than 36 sq. in. in facial area. To be entitled to this name, the tile must withstand the standard tests for "Weatherproof."

Faience Tile—This designates a tile having an impervious glaze and a dense body which is formed while in its plastic state and made principally from clay and other ceramic materials.

The appearance of this tile is characterized by a rugged, individual, although artistic variation of the face and edges like that occurring in handicraft methods of forming and finishing in the plastic state.

This type of tile usually has a heavier coating of glaze than glazed interior tile, and is more subject to heavier deposits of the glaze at the edges of the tile. It is well adapted to and widely used for manufacturing special architectural shapes or designs having a glazed finish. It is also obtainable in a large variety of sizes, shapes and designs. The tiles of this type are:

Faience—Symbol FT. Tile at least ½″ thick, varying to greater thickness for the larger sizes. This tile is larger than 6 sq. in. in facial area, except in cases where smaller sizes are needed to complete a pattern with the larger sizes.

Faience Mosaics—Symbol FM. Tile between ¼″ and ⅜″ in thickness which is less than 6 sq. in. in facial area.

Weatherproof Faience—Symbol WF. Tile which withstands the "Weatherproof" test without disintegration and conforms to "Faience" as previously stated.

Weatherproof Faience Mosaics—Symbol WFM. Tile which withstands the "Weatherproof" test without disintegration and conforms to sizes stated for "Faience Mosaics."

UNGLAZED TILE

In general, unglazed tile is composed of the same ingredients throughout the entire body (mass) as appear on the face of the tile. The ingredients of the body may be clay, shale, or other ceramic materials, and manufactured by various methods. The surface may be either plain, mottled, or "fire-flashed."

Ceramic Mosaic—This designates certain standard sizes of unglazed tile which are less than 6 sq. in. in facial area and approximately ¼″ to ⅜″ thick. Ceramic mosaic is usually pasted on sheets of paper to assist in setting. It is usually arranged in a mosaic pattern formed by the size or color of the tile.

Paver—This name designates certain sizes of unglazed tile (quarry and hand-made unglazed tile excluded). It is 6 sq. in. or more in facial area, except smaller sizes, which may be required to complete a pattern of the larger sizes. Unless specifically stated otherwise, pavers are ½ " thick (minimum ⅜ " thick) varying to a greater thickness for the larger sizes. These sizes are usually shipped unattached to paper and laid individually.

Unglazed tile having either an impervious, vitreous, or semi-vitreous body is suitable for interior or exterior use subject to freezing temperatures in the presence of water.

Porcelain Tile—This designates a vitreous unglazed tile made from clay and other ceramic materials refined and usually intimately mixed by washing. Tile of this type is characterized by a fine-grained dense body, sharply and precisely formed edges, and a smooth face impervious to stains. It is easily cleaned.

The color of the body, except white, is secured principally by the admixture of mineral oxides and stains. This tile is available in white, black and various colors, having either a plain, mottled or flashed face. The color effects, excepting red, are generally brighter, clearer and purer than those of other types of unglazed tile. If white or very light colors of unglazed tile are desired, specify "porcelain" type so as to be sure of the nonstaining qualities of the lighter colors. The names of this type of tile are:

Porcelain Ceramic Mosaic—Symbol PCM. Designating tile within the ceramic mosaic sizes described.

Porcelain Pavers—Symbol PP. Designating tile within the paver sizes described.

Natural Clay Tile—This designates unglazed vitreous or semi-vitreous tile made principally from clays and shales, and to lesser degree, other ceramic materials which are usually mixed together without washing and are formed by either pugging, cutting, or pressing.

Tile of this type is characterized by precisely formed face and edges, and a dense body with a granular structure. The face of the tile presents a somewhat rugged surface, which is rough enough to minimize slipping and yet sufficiently smooth to be readily cleaned.

The color of the body is secured principally from the natural firing colors of the clays. This tile is available in various colors, but principally in black, reds, browns, tans, grays or blends thereof having either a plain, mottled or flashed face. The names of this type of tile are:

Natural Clay Ceramic Mosaic—Symbol NCM. Designating tile within the ceramic mosaic sizes described.

Natural Clay Pavers—Symbol NCP. Designating tile generally over 6 sq. in. in facial area and otherwise within the paver sizes described.

Abrasive Tile—This designates an unglazed tile of either the porcelain of natural clay type with an admixture of approximately 5% by weight of abrasive grain (such as carborundum or alundum).

By reason of the abrasive grain, tile of this type is characterized by a coarse, rugged face, and are extremely resistant to slipping. The face and edges of the tile are rugged but reasonably straight. The abrasive grain is thoroughly impregnated within the body of the tile. The surface can be readily cleaned. This tile presents a speckled effect resulting from the interspersed abrasive grain showing on the surface of the tile. The names of this type of tile are:

Abrasive Ceramic Mosaic—Symbol ACM. Designating tile within the ceramic mosaic sizes described.

Abrasive Paver—Symbol AP. Designating tile within the paver size described.

Glazed Ceramic Mosaic Tile—Symbol GCM. This designates a tile within the limits of the sizes of ceramic mosaic.

It may be glazed with either a transparent, white, black or colored glaze, or with polychrome and mottled effect. The body may be either the porcelain or the natural clay type and is usually the general color of the glaze applied.

Quarry Tile—Symbol QT. This designates an unglazed tile made entirely or principally from clay and shale. It is usually formed in the plastic state by extrusion or pressing. Tile of this type is characterized by a dense body and usually has the somewhat rugged appearing face which is characteristic of larger sized tile mechanically formed in the plastic state. However, the face and edges are reasonably straight and smooth, and are readily cleaned.

The color is secured principally from the natural firing colors of the clay and shale used. The colors are principally red, buff, gray, and blends thereof, with either a plain or a flashed face. This tile is, unless specifically stated otherwise, a minimum of ½ ″ thick, varying to greater thickness to meet manufacturing convenience. It is generally larger in facial area than 6 sq. in., but smaller sizes may be furnished as part of a pattern of the larger sizes.

Hand-Made Unglazed Tile—Symbol HMU. This designates an unglazed tile which is formed while in its plastic state. The body is composed principally of clay, and to a lesser degree, other ceramic materials, prepared or blended by individually convenient methods. The appearance of this tile is characterized by a rugged individual although artistic variation of face and edges like that occurring in handicraft methods of forming and finishing in the plastic state.

The tile is a minimum of ½ ″ thick, varying to greater thickness to meet manufacturing convenience. It is generally larger in facial area than 6 sq. in., but tile of smaller size is furnished if needed as part of a pattern of larger sizes. This type of body is adapted to, and widely used for, manufacturing special architectural shapes or designs having an unglazed finish.

CHAPTER 16

Terrazzo

Terrazzo is derived from the Italian word "terrace" and by definition: It is a form of mosaic flooring made by embedding small pieces of marble in mortar and polishing.

Today, this traditional material is obtained by combining marble chips of selected size and texture, in a matrix of portland cement or the more recent synthetic or resinous matrices.

Terrazzo is a composition material, poured in place or precast, which is used for floor and wall construction. It consists of marble chips, seeded or unseeded, with a binder added that is cementitious, noncementitious, or a combination of both. The terrazzo is poured into forms, cured and then ground and polished.

MARBLE CHIPS

Marble of various types and colors is quarried, carefully selected to avoid off-color or contaminated material. It is crushed by a process that will largely eliminate flat or slivery chips and accurately sized to yield marble chips for the terrazzo floor. Marble is sometimes defined as a metamorphic rock formed by the recrystallization of limestone. However, for centuries from the commercial standpoint, and in recent decades from a geological standpoint, marble has been redefined to include all calcareous rocks capable of taking a polish. Excellent domestic and imported marble chips are available for use in terrazzo in a wide range of colors, which may be combined in infinite variety to meet the exacting color harmonies demanded in the finish of structures of every description.

Marble Chip Sizes

Marble chips are graded by number according to size in accordance with standards adopted by producers, as shown in Table 1.

Table 1. Grading of Marble Chips According to Size

Number	Passes Screen (in inches)	Retained on Screen (in inches)
0	1/8	1/16
1	1/4	1/8
2	3/8	1/4
3	1/2	3/8
4	5/8	1/2
5	3/4	5/8
6	7/8	3/4
7	1	7/8
8	1 1/8	1

CEMENT

Both gray and white portland cement are widely used in the installation of terrazzo floors. These cements are identical in structural properties (ASTM Specification C-150 Type 1 portland cement: Federal Specification SS-C-196 Type 1 portland cement). While white portland cement is more costly, due to the more careful selection of raw materials and greater color control during the processing and manufacturing, its use offers a clearer background for presentation of the colored marble chips where this is a desirable architectural treatment, and further lends itself better to treatment with coloring pigments. Since portland cement is generally obtained from plants nearest the project location, there might be slight variations between portland cements from different sources of manufacture, particularly in comparing gray portland cements from different areas of the United States.

DIVIDER STRIPS

The purpose of divider strips in terrazzo is to control and localize shrinkage and flexure cracks as much as possible. Although all known techniques are employed to eliminate shrinkage cracks, they often occur in spite of divider strips being correctly utilized and the observation of all other precautions. Divider strips are also used to permit the laying of different colors in adjoining panels and the use of designs and patterns to meet any aesthetic requirements.

Other Uses for Divider Strips

To break up large expanses of surface area, thus eliminating waviness and the illusion. This is particularly true in the use of monolithic and thin set terrazzo where strips are not a standard requirement.

To allow movement at all expansion and construction joints. It is strongly recommended that utilization is made of the neoprene filler expansion type strip to control joints. This strip has proven a success as a perimeter expansion strip in both large and small areas of monolithic terrazzo.

To allow the precise accuracy necessary to install lettering and special designs.

To terminate the height of poured-in-place base or wainscot, thus, providing an exact line for plaster or other material to finish against.

To form a neat transition from terrazzo to other flooring.

To divide poured in place cove base vertically at regular intervals. Divider stips are available in various heights and usually with a 1″ or 1½″ radius.

To attempt to anticipate the location of cracks in the concrete slab where monolithic terrazzo is to be used, and thus place divider strips at these vital positions.

There are many uses for strip not mentioned above which are mainly for aesthetic purposes, all of which emphasize the freedom and flexibility of terrazzo. The use of divider strips does not in any way affect the structural soundness of terrazzo.

COLOR PIGMENTS

The color pigments must be commercially pure. Natural or synthetic mineral oxides or other coloring materials manufactured for use in portland cement mixtures have proven to be satisfactory. Table 2 may be used as a guide to the approximate quantities of high-grade pigments required for the colors and shades indicated.

MIXTURES

The base for terrazzo finish is mixed in the proportions of 1 part portland cement to 4 parts clean coarse sand. The terrazzo mixture is in the proportions of 200 lbs. of aggregate to 1 sack of portland cement with not more than 4 gallons of water and the proper amount of pigment to produce the approved color. The cement and pigment is mixed dry to a uniform color before adding the other materials.

The terrazzo mixture should be of the driest consistency possible to work into place with a sawing motion of the strike-off board or straightedge. Changes in consistency are obtained only by changes in the proportions of aggregate and cement. *In no case shall the specified amount of mixing water be exceeded.*

PLACING

Method I—Bonded Finish

The surface of the structural base slab must be cleaned of all plaster and other materials that would interfere with the bond and must be thoroughly wetted. It is then slushed with a neat cement grout and thoroughly broomed into the surface. The underbed is then spread uniformly and brought to a level not less than ½ " nor more than ¾ " below the finished floor.

Method II—Broken Bond Finish

The surface of the structural base slab is covered with a uniform layer of fine sand ¼ " thick and covered with an approved tarpaper overlapping at least 2" at all edges. The underbed is then

341

Table 2. Pigments for Colored Concrete Floor Finish

Color desired	Commercial names of colors for use with cement	Approximate quantities required—lb. per bag of cement	
		Light shade	Medium shade
Grays, blue-black and black	Germantown lampblack* or carbon black* or black oxide of manganese* or mineral black	½ ½ 1 1	1 1 2 2
Blue	Ultramarine blue	5	9
Brownish red to dull brick red	Red oxide of iron	5	9
Bright red to vermilion	Mineral turkey red	5	9
Red sandstone to purplish red	Indian red	5	9
Brown to reddish-brown	Metallic brown (oxide)	5	9
Buff, colonial tint and yellow	Yellow ochre or yellow oxide	5 2	9 4
Green	Chromium oxide or greenish blue ultra-marine	5 6	9

*Only first-quality lampblack should be used. Carbon black is light in weight and requires very thorough mixing. Black oxide or mineral black is probably most advantageous for general use. For black, use 11 lbs. of oxide for each bag of cement.

spread uniformly and brought to a level not less than ½ " nor more than ¾ " below the finished floor.

While the underbed is in a semiplastic state, the dividing strips are installed to conform to the designs shown on the drawings. The top of the strips should be at least 1/32" above the finished level of the floor.

The terrazzo mix is placed in the spaces formed by the dividing strips and rolled into a compact mass by means of heavy rollers, adding aggregate if necessary so that the finished surface will show a minimum of 70% aggregate. Immediately after rolling, the surface should be floated and troweled to an even surface disclosing the lines of the strips on a level with the terrazzo filling.

CURING AND PROTECTION

All freshly placed concrete must be protected from the elements and from all defacements due to building operations. As soon as the concrete has hardened sufficiently to prevent damage, it is covered with at least 1" of wet sand or other covering satisfactory to the architect, and is kept continually wet by sprinkling with water for at least 7 days when using standard portland cement and for at least 3 days when using high early strength portland cement.

The temperature of the concrete at time of placing should be above 70°F and maintained above 70°F for at least 3 days or above 50°F for at least 5 days when using standard portland cement and above 70°F for at least 2 days or above 50°F for at least 3 days when using standard high-early-strength portland cement.

SURFACING

When the terrazzo concrete has hardened enough to prevent dislodgment of aggregate particles, it is machine rubbed using No. 24 grit abrasive stones for the initial rubbing and No. 80 grit abrasive stones for the second rubbing. The floor is kept wet during the rubbing process. All material ground off is removed by squeegeeing and flushing with water.

A grout of portland cement, pigment, and water, of the same kind and color as the matrix is applied to the surface, filling all voids. In not less than 72 hours after grouting, the grouting coat is removed and the surface polished to a satisfactory finish by machines using stones not coarser than No. 80 grit. After removing all loose material, the finish is scrubbed with warm water and soft soap and then mopped dry. Figs. 1 and 2 illustrate the surface texture of a light and dark color terrazzo floor.

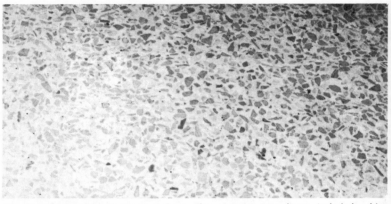

Fig. 1. A beige-colored terrazzo floor. The concrete ingredients included white portland cement and yellow oxide pigment. Pieces of various colored marble are embedded in the concrete.

Fig. 2. This floor is dark green with lighter-colored marble chips.

CARE OF TERRAZZO

It is doubtful that there is a flooring material in use today that requires less care than terrazzo. Yet, a number of people have difficulty in maintaining it. Like other materials, there are inherent properties of terrazzo that should be understood—once understood, maintenance problems are eliminated and the full beauty as well as economy is realized.

To best understand terrazzo is to first break it down into components—marble and portland cement. They are mixed together in a ratio of 2 parts marble and 1 part cement. During its installation additional marble is sprinkled on the surface so that a minimum of 70% marble shows on the finished floor.

When dealing with terrazzo, the use of a pure surface coating is unnecessary and ordinarily not recommended. The terrazzo surface is at least 70% marble and marble has a very low porosity. It absorbs very little liquid and most staining materials have too thick a consistency to become absorbed. That portion of the terrazzo that needs protection is, therefore, the portland cement; which is porous and will absorb stains. Pure surface waxes will protect—but are easily walked off, and will tend to make any smooth surface slippery. Terrazzo does not need protection from wear, it needs protection from absorption and this is achieved through the use of a penetrating sealer which is absorbed into the portland cement, sealing off its pores.

Internal Protection

Proper protection for terrazzo is accomplished internally rather than on the surface. Being internal, the process of waxing and rewaxing is eliminated and only periodic resealings are necessary. It is wise to note here that purely surface protection holds dirt and adds to your cleaning and can possibly present a stripping problem.

There are few natural products as beautiful as marble and shoe leather is one of the finest polishers or abrasives you can use. If you maintain a surface coating over marble, you will not allow the floor to take its natural sheen, so long identified with terrazzo floors.

Neutral Cleansers Only

Terrazzo should be cleaned only with neutral liquid cleaners. The cleaning cycle to be set up will be regulated by the amount of traffic. For general cleaning, one cup of neutral cleaner is mixed with each 3 gallons of water. For extremely dirty areas, increase the amount of cleaner. Wet mop the solution onto the floor, allow several minutes for the grime-dissolving action to take place, then squeegee, wet vacuum, or mop up the dirt laden solution. It is important that the floors be kept wet at all times during the cleaning operation to prevent dissolved soil from drying back onto the floor. Also, if the solution is mopped up, it is important that the custodian change his rinse water regularly so that complete removal is assured and unsightly moplines are eliminated.

Electric scrubbing machines used periodically with a solution of neutral cleaner will loosen dirt that is hard to remove during normal daily wet mop cleaning.

The liquid cleaner selected must be neutral with a *pH* of less than 10.0% and free from any harmful alkali, acid, etc. that may ruin the floor. The N.T.M.A. specifically warns that soaps and scrubbing powders containing water solubles, inorganic salts or crystallizing salts should never be used in the maintenance of terrazzo. Many terrazzo, quarry, and ceramic tile floors have been destroyed by improper selection of cleaning materials.

Non-Oily Dressings

If a mop dressing is used for daily sweeping, be sure it is non-oily. Sweeping compounds containing oil are a fire hazard and most of them contain sand, which is hard to sweep up and abrades if left on the floor. Floor oils are not only a fire hazard but they will penetrate and permanently discolor terrazzo.

We do not know of any flooring material that, when under constant use, will not begin to show some staining from daily use and abuse—fruit, chewing gum, soda pop, cigarettes, on and on the list could go. In a carefully maintained building the custodians should be instructed to treat stains as soon as possible, as they become more difficult to remove after they have dried. However, no one should attempt to remove a stain until he knows what the stain is and why a certain type of remover is being used. Only as a last resort should chemicals be used to remove stains.

Stain Removal

Environment has a definite effect upon the quality of an employee's work as well as his reaction to his supervisor and the general public. The appearance of your floors, free from stains, will do much to achieve this.

Stain removers either dissolve the substance that causes the stain; absorb the stain; or act as a bleaching agent. Thus, stain removers fall into three general classes:

1. Solvents such as carbon tetrachloride which dissolve grease, chewing gum, lipstick, etc.
2. Absorbents such as chalk, talcum powder, blotting paper or cotton which absorb fresh grease or moist stains.
3. Bleaches such as household ammonia, hydrogen peroxide, acetic acid or lemon juice which discolor stains.

In removing stains with chemicals, directions should be carefully followed. For example, when the procedure specifies treatment with a solvent before cleaning, it may be that if this was reversed the alkali in the soap would set the stain and make it impossible to remove.

Appreciate the importance of knowing the surface to be treated and the nature of the stain before trying to remove it. The housekeeper should ask these questions:

1. Is it a water base stain? If so, water will remove it.
2. Is the stain alcohol? If so, alcohol will remove it (for example, tincture of iodine).
3. Is the stain acid? If so, use an alkali to neutralize.
4. Is the stain alkali? Then use acid to neutralize it.
5. Is the stain grease? If so, use soap.
6. Does the stain contain albumin, as in milk or blood? If so, do not use a hot solution. It will cook the albumin.

For common stains in your building, prepare a chart listing the kind of remover to use on specific stains or specific surfaces. It is a constant challenge to produce effective results. Remember, floors (any floors) that are deeply embedded with sand or soil deteriorate quickly and, once they have started wearing, no amount of cleaning will bring them back.

Index

AUDEL BOOKS *practical reading for profit*

APPLIANCES

Air Conditioning (23159)

Domestic, commercial, and automobile air conditioning fully explained in easily-understood language. Troubleshooting charts aid in making diagnosis and repair of system troubles.

Commercial Refrigeration (23195)

Installation, operation, and repair of commercial refrigeration systems. Included are ice-making plants, locker plants, grocery and supermarket refrigerated display cases, etc. Trouble charts aid in the diagnosis and repair of defective systems.

Air Conditioning and Refrigeration Library—2 Vols. (23196)

Home Appliance Servicing—3rd Edition (23214)

A practical "How-To-Do-It" book for electric & gas servicemen, mechanics & dealers. Covers principles, servicing and repairing of home appliances. Tells how to locate troubles, make repairs, reassemble and connect, wiring diagrams and testing methods. Tells how to fix electric refrigerators, washers, ranges, toasters, ironers, broilers, dryers, vacuum sweepers, fans, and other appliances.

Home Refrigeration and Air Conditioning (23133)

Covers basic principles, servicing, operation, and repair of modern household refrigerators and air conditioners. Automotive air conditioners are also included. Troubleshooting charts aid in trouble diagnosis. **A gold mine of essential facts for engineers, servicemen, and users.**

AUTOMOTIVE

Automobile Guide (23192)

New revised edition. Practical reference for auto mechanics, servicemen, trainees, and owners. Explains theory, construction, and servicing of modern domestic motorcars. FEATURES: All parts of an automobile—engines—pistons—rings—connecting rods—crankshafts—valves—cams—timing—cooling systems—Fuel-feed systems—carbureators — automatic choke — transmissions — clutches — universals — propeller shafts—dierentials—rear axles—running gear—brakes—wheel alignment—steering gear—tires—lubrication—ignition systems—generators and alternators—starters—lighting systems—batteries—air conditioning—cruise controls—emission control systems.

Auto Engine Tune-up (23181)

New revised edition. This popular how-to-do-it guide shows exactly how to tune your car engine for extra power, gas economy, and fewer costly repairs. New emission-control systems are explained along with the proper methods for correcting faults and making adjustments to keep these systems in top operating condition.

Automotive Library—2 Vols. (23198)

Diesel Engine Manual (23199)

A practical treatise on the theory, operation and maintenance of modern Diesel engines. Explains Diesel principles—valves—timing—fuel pumps—pistons and rings—cylinders—lubrication—cooling system—fuel oil—engine indicator—governors—engine reversing—answers on operation—calculations. AN IMPORTANT GUIDE FOR ENGINEERS, OPERATORS, STUDENTS.

Gas Engine Manual   (23061)

A completely practical book covering the construction, operation and repair of all types of modern gas engines. Part I covers gas-engine principles; engine parts; auxiliaries; timing methods; ignition systems. Part II covers troubleshootng, adjustment and repairs.

Auto Body Repair for the Do-it-yourselfer (23238)

Another popular Audel paper back book covering auto body repair and body maintenance. This book shows the do-it-yourself car owner how to save money by following the easy detailed step-by-step procedure of how to use touch-up paint; how to prevent rust from spreading; how to repair rust spots and dents by using various body fillers. It also covers in great detail the repair of vinyl tops, cleaning upholstery and removing scratches from window glass.

BUILDING AND MAINTENANCE

Answers on Blueprint Reading   (23041)

Covers all types of blueprint reading for mechanics and builders. The man who can read blueprints is in line for a better job. This book gives you the secret language, step by step in easy stages. NO OTHER TRADE BOOK LIKE IT.

Building Construction and Design   (23180)

A completely revised and rewritten version of Audel's **Architects and Builders Guide.** New illustrations and extended coverage of material makes this treatment of the subject more valuable than ever. Anyone connected in any way with the building industry will profit from the information contained in this book.

Building Maintenance   (23140)

A comprehensive book on the practical aspects of building maintenance. Chapters are included on: painting and decorating; plumbing and pipe fitting; carpentry; calking and glazing; concrete and masonry; roofing; sheet metal; electrical maintenance; air conditioning and refrigeration; insect and rodent control; heating maintenance management; cutodial practices: A BOOK FOR BUILDING OWNERS, MANAGERS, AND MAINTENANCE PERSONNEL.

Gardening & Landscaping   (23229)

A comprehensive guide for the homeowner, industrial, municipal, and estate grounds-keepers. Information on proper care of annual and perennial flowers; various house plants; greenhouse design and construction; insect and rodent control; complete lawn care; shrubs and trees; and maintenance of walks, roads, and traffic areas. Various types of maintenance equipment are also discussed.

Carpenters & Builders Library—4 Vols.   (23244)

A practical illustrated trade assistant on modern construction for carpenters, builders, and all woodworkers. Explains in practical, concise language and illustrations all the principles, advances and short cuts based on modern practice. How to calculate various jobs.

Vol. 1—**(23240)**—Tools, steel square, saw filing, joinery, cabinets.
Vol. 2—**(23241)**—Mathematics, plans, specifications, estimates.
Vol. 3—**(23242)**—House and roof framing, laying out, foundations.
Vol. 4—**(23243)**—Doors, windows, stairs, millwork, painting.

Carpentry and Building   (23142)

Answers to the problems encountered in today's building trades. The actual questions asked of an architect by carpenters and builders are answered in this book. No apprentice or journeyman carpenter should be without the help this book can offer.

Do-It-Yourself Encyclopedia   (23207)

An all-in-one home repair and project guide for all do-it-yourselfers. Packed with step-by-step plans, thousands of photos, helpful charts. A really authentic, truly monumental, home-repair and home-project guide.

Home Workshop & Tool Handy Book   (23208)

The most modern, up-to-date manual ever designed for home craftsmen and do-it-yourselfers. Tells how to set up your own home workshop, (basement, garage, or spare room), all about the various hand and power tools (when, where, and how to use them, etc.). Covers both wood- and metal-working principles and practices. An all-in-one workshop guide for handy men, professionals and students.

Plumbers and Pipe Fitters Library—3 Vols.   (23155)

A practical illustrated trade assistant and reference for master plumbers, journeyman and apprentice pipe fitters, gas fitters and helpers, builders, contractors, and engineers. Explains in simple language, illustrations, diagrams, charts, graphs and pictures, the principles of modern plumbing and pipe-fitting practices.

Vol. 1—(23152)—Materials, tools, calculations
Vol. 2—(23153)—Drainage, fittings, fixtures.
Vol. 3—(23154)—Installation, heating, welding

Home Plumbing Handbook (23239)

A paper back book filled with valuable information on faucet construction and repair—repairing water closet tanks—what to do about stopped up drains and repair or replacing automatic water heaters. There is a lot of valuable information on installing garbage disposers, dishwashers, homemade humidifiers and how to remodel a kitchen and bathroom. Also included are full chapters on septic tanks and disposal fields, private water systems and water conditioning.

Masons and Builders Library—2 Vols.   (23185)

A **practical illustrated trade assistant on modern construction for bricklayers, stonemasons, cement workers, plasterers, and tile setters.** Explains in clear language and with detailed illustrations all the principles, advances, and shortcuts based on modern practice—including how to figure and calculate various jobs.

Vol. 1—(23182)—Concrete, Block, Tile, Terrazzo.
Vol. 2—(23183)—Bricklaying, Plastering, Rock Masonry, Clay Tile.

Upholstering   (23189)

Upholstering is explained for the average householder and apprentice upholsterer in this Audel text. Selection of coverings, stuffings, springs, and other upholstering material is made simple. From repairing and regluing of the bare frame, to the final sewing or tacking, for antiques and most modern pieces, this book gives complete and clearly written instructions and numerous illustrations.

Questions and Answers for Plumbers Examinations   (23206)

Many questions are answered as to types of fixtures to use, size of pipe to install, design of systems, size and location of septic tank systems, and procedures used in installing material. Subjects such as traps, cleanouts, drainage, vents, water supply and distribution, sewage treatment, plastic pipe, steam and hot water fittings, just to name a few.

Wood Furniture, Finishing, Refinishing Repairing   (23216)

The basic technical and practical information needed for complete wood finishing. This book presents the fundamentals of furniture repair, both veneer and solid wood and complete refinishing procedures, which includes stripping the old finish, sanding, selecting the finish and using wood fillers. Complete step-by-step procedure for antiquing, painting, staining, flocking, inlay patterns and gold- and silver-leaf finishing. Various woods, along with actual grain photos, are discussed as to their characteristics and their reactions to various finishes.

Building a Vacation Home: Step-by-Step   (23222)

Step-by-step procedure in the construction of this vacation home includes fifteen chapters, numerous illustrations, actual photographs, and a complete set of blueprints, showing cabinet construction as well as building and foundation structure. It points out difficulties which might be avoided by informing you of the proper procedures for obtaining property lines, building and sewage permits, and various hidden costs which are bound to occur.

Oil Burners   (23151)

Provides complete information on all types of oil burners and associated equipment. Discusses burners—blowers—ignition transformers—electrodes—nozzles—fuel pumps—filters controls. Installation and maintenance are stressed. Trouble shooting charts permit rapid diagnosis of system troubles and possible remedies t correct them.

Heating Ventilating and Air Conditioning Library   (2322

An introduction to the fundamentals of installing, servicing, and repairing vario types of equipment used in residential heating, ventilation, and air conditioning sy tems. All subjects are covered; from heating fundamentals, types of heating equi ment, heating controls, air conditioning equipment, humidity, and air cleaners ar filters.

> Vol. 1—(23248)   Heat fundamentals—heating systems—insulation principles
> Vol. 2—(23249)   Burners—controls—ducts—valves
> Vol. 3—(23250)   Radiant heat—air conditioning—air cleaners and filters—
>   humidifiers—dehumidifiers

ELECTRICITY-ELECTRONICS

Wiring Diagrams for Light & Power   (23232)

Electricians, wiremen, linemen, plant superintendents, construction engineers, ele trical contractors, and students will find these diagrams a valuable source of practic help. Each diagram is complete and self-explaining. A PRACTICAL HANDY BOO OF ELECTRICAL HOOK-UPS.

Electric Motors   (23150)

Covers the construction, theory of operation, connection, control, maintenance, ar troubleshooting of all types of electric motors. A HANDY GUIDE FOR ELECTRICIAN AND ALL ELECTRICAL WORKERS.

Practical Electricity   (23218)

This updated version is a ready reference book, giving complete instruction ar practical information on the rules and laws of electricity—maintenance of electric machinery—AC and DC motors—wiring diagrams—lighting—house and power wirir —meter and instrument connections—transformer connectors—circuit breakers— power stations—automatic substations. THE KEY TO A PRACTICAL UNDERSTANDIN OF ELECTRICITY.

House Wiring—3rd Edition   (23224)

Answers many questions in plain simple language concerning all phases of hou wiring. A ready reference book with over 100 illustrations and concise interpretatio of many rulings contained in the National Electrical Code. Electrical contracto wiremen, and electricians will find this book invaluable as a tool in the electric field

Guide to the 1975 National Electric Code   (23223)

This important and informative book is now revised to conform to the 1975 Nation Electrical Code. Offers an interpretation and simplification of the rulings contain in the Code. Electrical contractors, wiremen, and electricians will find this bo invaluable for a more complete understanding of the NEC.

Questions and Answers for Electricians Examinations (23225)

Newly revised to conform to the 1975 National Electrical Code. A practical book to help you prepare for all grades of electricians examinations. A helpful review of fundamental principles underlying each question and answer needed to prepare you to solve any new or similar problem. Covers the NEC; questions and answers for license tests; Ohm's law with applied examples, hook-ups for motors, lighting, and instruments. A COMPLETE REVIEW FOR ALL ELECTRICAL WORKERS.

Electrical Library—6 Vols. (23236)

Electric Generating Systems (23179)

Answers many questions concerning the selection, installation, operation, and maintenance of engine-driven electric generating systems for emergency, standby, and away-from-the-power-line applications. Private homes, hospitals, radio and television stations, and pleasure boats are only a few of the installations that owners either desire or require for primary power or for standby use in case of commercial power failure. THE MOST COMPREHENSIVE COVERAGE OF THIS SUBJECT TO BE FOUND TODAY.

Electrical Course for Apprentices and Journeymen (23209)

A basic study course for apprentice or journeyman electricians which may be used as a classroom or self-taught program. This book can be utilized without any other books on electrical theory. Review questions included.

Electronic Security Systems (23205)

Protect your home and business. A basic and practical text written for the electrician, electronic technician, security director and do-it-yourself householder or businessman. Such subjects as sensors and encoders, indicators and alarms, electrical control and alarm circuits, security communications, closed circuit television, and security system installations are covered.

ENGINEERS-MECHANICS-MACHINISTS

Machinists Library (23174)

Covers modern machine-shop practice. Tells how to set up and operate lathes, screw and milling machines, shapers, drill presses and all other machine tools. A complete reference library. A SHOP COMPANION THAT ANSWERS YOUR QUESTIONS.

Vol. 1—(23175)—Basic Machine Shop.
Vol. 2—(23176)—Machine Shop.
Vol. 3—(23177)—Toolmakers Handy Book.

Millwrights and Mechanics Guide— 2nd Edition (23201)

Practical information on plant installation, operation, and maintenance for millwrights, mechanics, maintenance men, erectors, riggers, foremen, inspectors, and superintendents. Partial contents: • Drawing and Sketching • Machinery Installation • Power-Transmission Equipment • Couplings • Packing and Seals • Bearings • Structural Steel • Mechanical Fasteners • Pipe Fittings and Valves • Carpentry • Sheet-Metal Work • Blacksmithing • Rigging • Electricity • Welding • Mathematics and much more.

Practical Guide to Mechanics (23102)

A convenient reference book valuable for its practical and concise explanations of the applicable laws of physics. Presents all the basics of mechanics in everyday language, illustrated with practical examples of their applications in various fields.

Power Plant Engineers Guide—2nd Edition   (23220)

A complete steam or diesel power plant engineers library. This book covers (in question-and-answer form) facts for all engineers, fireman, water tenders, oilers, and operators of steam and diesel power systems. A valuable book to the applicant for engineer's and fireman's licenses.

Questions & Answers for Engineers and Firemans Examinations   (23217)

An aid for stationary, marine, diesel & hoisting engineers' examinations for all grades of licenses. A new concise review explaining in detail the principles, facts and figures of practical engineering. Questions & Answers.

Welders Guide—2nd Edition   (23202)

New revised edition. Covers principles of electric, oxyacetylene, thermit, unionmelt welding for sheet metal; spot and pipe welds; pressure vessels; aluminum, copper brass, bronze, plastics, and other metals; airplane work; surface hardening and hard facing; cutting brazing; underwater welding; eye protection. EVERY WELDER SHOULD OWN THIS GUIDE.

Mechanical Trades Pocket Manual   (23215)

This "paper back—pocket size manual" provides reference material for mechanical tradesman. The handbook covers methods, tools, equipment, procedures, etc. in convenient form and plain language to aid the mechanic in performance of day-to-day tasks concerned with installation, maintenance, and repair of machinery and equipment.

FLUID POWER

Pumps   (23167)

A detailed book on all types of pumps from the old-fashioned kitchen variety to the most modern types. Covers construction, application, installation, and troubleshooting.

MATHEMATICS

Practical Mathematics for Everyone—2 Vols.   (23112)

A concise and reliable guide to the understanding of practical mathematics. People from all walks of life, young and old alike, will find the information contained in these two books just what they have been looking for. The mathematics discussed is for the everyday problems that arise in every household and business.
Vol. 1—(23110)—Basic Mathematics.
Vol. 2—(23111)—Financial Mathematics.

OUTBOARD MOTORS

Outboard Motors & Boating   (23168)

Provides the information necessary to adjust, repair, and maintain all types of outboard motors. Valuable information concerning boating rules and regulations is also included.

RADIO-TELEVISION-AUDIO

Handbook of Commercial Sound Installations   (23126)

A practical complete guide to planning commercial systems, selecting the most suitable equipment, and following through with the most proficient servicing methods. For technicians and the professional and businessman interested in installing a sound system.

Practical Guide to Auto Radio Repair   (23128)

A complete servicing guide for all types of auto radios, including hybrid, all-transistor, and FM . . . PLUS removal instructions for all late model radios. Fully illustrated.

Practical Guide to Servicing Electronic Organs   (23132)

Detailed, illustrated discussions of the operation and servicing of electronic organs. Including models by Allen, Baldwin, Conn, Hammond, Kinsman, Lowrey, Magnavox, Thomas, and Wurlitzer.

Radioman's Guide   (23163)

Audel best-seller, containing the latest information on radio and electronics from the basics through transistors. Covers radio fundamentals—Ohm's law—physics of sound as related to radio—radio-wave transmission—test equipment—power supplies—resistors, inductors, and capacitors—transformers—vacuum tubes—transistors—speakers—antennas—troubleshooting. A complete guide and a perfect preliminary to the study of television servicing.

Television Service Manual   (23162)

Includes the latest designs and information. Thoroughly covers television with transmitter theory, antenna designs, receiver circuit operation and the picture tube. Provides the practical information necessary for accurate diagnosis and repair of both black-and-white and color television receivers. A MUST BOOK FOR ANYONE IN TELEVISION.

Radio-TV Library—2 Vol.   (23161)

SHEET METAL

Sheet Metal Workers Handy Book   (23046)

Containing practical information and important facts and figures. Easy to understand. Fundamentals of sheet metal layout work. Clearly written in everyday language. Ready reference index.

TO ORDER AUDEL BOOKS mail this handy form to

Theo. Audel & Co., 4300 W. 62nd
Indianapolis, Indiana 46268

Please send me for FREE EXAMINATION books marked (x) below. If I de-
cide to keep them I agree to mail $3 in 10 days on each book or set ordered
and further mail ⅓ of the total purchase price 30 days later, with the
balance plus shipping costs to be mailed within another 30 days. Otherwise, I will return them for refund.

APPLIANCES

- (23196) Air Conditioning & Refrigeration Library (2 Vols.) 13.75
  - (23159) Air Conditioning 6.95
  - (23195) Commercial Refrigeration 7.50
- (23214) Home Appliance Servicing 7.95
- (23133) Home Refrigeration and Air Conditioning 8.95
- (23151) Oil Burners 6.25

AUTOMOTIVE

- (23198) Automotive Library (2 Vols.) 15.95
  - (23192) Automotive Guide 10.25
  - (23181) Auto Engine Tune-Up 6.75
- (23199) Diesel Engine Manual 8.50
- (23061) Gas Engine Manual 6.50
- (23238) Auto Body Repair for Do-it-yourselfer 5.95

BUILDING AND MAINTENANCE

- 923041) Answers on Blueprint Reading 6.50
- (23180) Building Construction and Design 6.75
- (23140) Building Maintenance 7.50
- (23244) Carpenters and Builders Library (4 Vols.) 22.50
  - Single Volumes sold separately .... ea. 5.95
- (23142) Carpentry and Building 7.50
- (23207) Do-It-Yourself Encyclopedia 13.50
- (23229) Gardening & Landscaping 7.95
- (23208) Home Workshop & Tool Handy Book 6.50
- (23185) Masons & Builders Library (2 Vols.) 12.95
  - Single Volumes sold separately .... ea. 6.95
- (23189) Upholstering 6.75
- (23155) Plumbers and Pipe Fitters Library (3 Vols.) 17.00
  - Single Volumes sold separately .... ea. 6.00
- (23206) Q & A for Plumbers Exam 6.95
- (23239) Home Plumbing Handbook 8.95
- (23227) Heating, Ventilating and Air Conditioning (3 Vols.) 22.50
  - Single Volumes sold separately 7.95
- (23216) Wood Furniture Finishing, Repair 6.95
- (23222) Building a Vacation Home 7.95

ELECTRICITY-ELECTRONICS

- (23179) Electric Generating Systems 6.75
- (23236) Electrical Library (6 Vols.) 38.00

- (23232) Wiring Diagrams for Light and Power 6.95
- (23150) Electric Motors 6.95
- (23218) Practical Electricity 6.95
- (23224) House Wiring (3rd Edition) 6.50
- (23223) Guide to the 1975 National Electrical Code 8.50
- (23225) Questions & Answers for Electrician's Exams 6.50
- (23205) Electronic Security Systems 6.95
- (23209) Electrical Course for Apprentices & Jour-
  neymen 6.95

ENGINEERS-MECHANICS-MACHINISTS

- (23174) Machinists Library (3 Vols.) 19.50
  - Single Volumes sold separately ...... ea. 6.75
- (23202) Welders Guide (2nd Edition) 10.95
- (23215) Mechanical Trades Pocket Manual 3.95
- (23201) Milwrights and Mechanics Guide (2nd Edition) 10.95
- (23220) Power Plant Engineers Guide 9.95
- (23102) Practical Guide to Mechanics 5.50
- (23217) Q&A for Engineers and Firemans Exams 7.95

FLUID POMER

- (23167) Pumps 8.95

MATHEMATICS

- (23112) Practical Math for Everyone (2 Vols.) 10.25
  - Single Volumes sold separately ...... ea. 5.50

OUTBOARD MOTORS

- (23168) Outboard Motors and Boating 5.50

RADIO-TELEVISON-AUDIO

- (23126) Handbook of Commercial Sound Installations 5.95
- (23128) Practical Guide to Auto Radio Repair 4.95
- (23132) Practical Guide to Servicing Electronic Organs 5.50
- (23161) Radio & Television Library (2 Vols.) 13.75
  - (23163) Radiomans Guide 6.75
  - (23162) Television Service Manual 5.75

SHEET METAL

- (23046) Sheet Metal Workers Handy Book 5.75

Prices Subject to Change Without Notice

Name_____

Address_____

City_____ State _____ Zip _____

Occupation_____ Employed by _____

**SAVE SHIPPING CHARGES! Enclose Full Payment
With Coupon and We Pay Shipping Charges.** PRINTED IN U.S.A.